T0156014

Lecture Notes in Computer Science 14160

Founding Editors

Gerhard Goos
Juris Hartmanis

The series Lecture Notes in Computer Science (LNCS), including its subseries Lecture Notes in Artificial Intelligence (LNAI) and Lecture Notes in Bioinformatics (LNBI), has established itself as a medium for the publication of new developments in computer science and information technology research, teaching, and education.

LNCS enjoys close cooperation with the computer science R & D community, the series counts many renowned academics among its volume editors and paper authors, and collaborates with prestigious societies. Its mission is to serve this international community by providing an invaluable service, mainly focused on the publication of conference and workshop proceedings and postproceedings. LNCS commenced publication in 1973.

Abdelkader Hameurlain · A Min Tjoa ·
Omar Boucelma · Farouk Toumani
Editors

Transactions on Large-Scale Data- and Knowledge- Centered Systems LIV

Special Issue on Data Management - Principles, Technologies, and Applications

 Springer

Editors-in-Chief
Abdelkader Hameurlain
Paul Sabatier University, IRIT
Toulouse, France

A Min Tjoa
Technical University of Vienna, IFS
Vienna, Austria

Guest Editors
Omar Boucelma
Aix-Marseille University, LIS
Marseille, France

Farouk Toumani
Université Clermont Auvergne, CNRS,
LIMOS
Aubiere, France

ISSN 0302-9743　　　　　　ISSN 1611-3349 (electronic)
Lecture Notes in Computer Science
ISSN 1869-1994　　　　　　ISSN 2510-4942 (electronic)
Transactions on Large-Scale Data- and Knowledge-Centered Systems
ISBN 978-3-662-68013-1　　　ISBN 978-3-662-68014-8 (eBook)
https://doi.org/10.1007/978-3-662-68014-8

Preface

This volume features a selection of five papers from the 38th Conference on Data Management – Principles, Technologies, and Applications (BDA 2022).

Within this special issue, we selected three articles covering noteworthy research areas, including temporal graph management systems, time series prediction models and exploratory analysis, and data clustering and classification. Authors of the selected papers were invited to prepare and submit journal versions of their contributions, which underwent a thorough re-review by the editorial board of this issue.

Furthermore, we are pleased to present two additional extended papers from invited talks at BDA 2022. These papers address significant topics such as explanations for query answers in databases and the key challenges encountered in healthcare data analytics over knowledge graphs. Including these papers adds depth and relevance to the special issue, addressing critical topics within the field.

We sincerely thank the authors and reviewers who dedicated their time and expertise to making this special issue successful. Their hard work and valuable insights significantly shaped the content of this volume. We would also like to thank the Editors-in-Chief, Abdelkader Hameurlain and A Min Tjoa, for their guidance and support throughout this endeavor.

As part of the TLDKS journal series, this special issue aims to contribute to advancing data management research and its practical applications. We hope that the papers presented here inspire further exploration, spark new ideas, and ultimately contribute to the growth and development of the field.

July 2023

Omar Boucelma
Farouk Toumani

Organization

Editors-in-Chief

Abdelkader Hameurlain   Paul Sabatier University, IRIT, France
A Min Tjoa   Technical University of Vienna, IFS, Austria

Guest Editors

Omar Boucelma   Aix-Marseille University, LIS, France
Farouk Toumani   University Clermont Auvergne, CNRS, LIMOS, France

Editorial Board

Reza Akbarinia   Inria, France
Dagmar Auer   Johannes Kepler University Linz, Austria
Djamal Benslimane   University Lyon 1, France
Stéphane Bressan   National University of Singapore, Singapore
Mirel Cosulschi   University of Craiova, Romania
Johann Eder   Alpen Adria University of Klagenfurt, Austria
Anna Formica   National Research Council in Rome, Italy
Shahram Ghandeharizadeh   University of Southern California, USA
Anastasios Gounaris   Aristotle University of Thessaloniki, Greece
Sergio Ilarri   University of Zaragoza, Spain
Petar Jovanovic   Universitat Politècnica de Catalunya, BarcelonaTech, Spain
Aida Kamišalić Latifić   University of Maribor, Slovenia
Dieter Kranzlmüller   Ludwig-Maximilians-Universität München, Germany
Philippe Lamarre   INSA Lyon, France
Lenka Lhotská   Technical University of Prague, Czech Republic
Vladimir Marik   Technical University of Prague, Czech Republic
Jorge Martinez Gil   Software Competence Center Hagenberg, Austria
Franck Morvan   Paul Sabatier University, IRIT, France
Torben Bach Pedersen   Aalborg University, Denmark
Günther Pernul   University of Regensburg, Germany

Contents

Clock-G: Temporal Graph Management System

Maria Massri[1,2(✉)], Zoltan Miklos[1], Philippe Raipin[2], and Pierre Meye[2]

[1] Univ Rennes CNRS IRISA, Rennes, France
{maria.massri,zoltan.miklos}@irisa.fr
[2] Orange Labs, Cesson Sévigné, France
{maria.massri,philippe.raipin}@orange.com, meye_pierre@yahoo.fr

Abstract. Graphs are a ubiquitous data model for capturing entities
and their relationships. Since most graphs that model real-world networks evolve over time, efficiently managing temporal graphs is an important problem from both a theoretical and practical perspective. Querying the history of temporal graphs can lead to new applications such
as object tracking, anomaly detection, and predicting future behavior.
However, existing commercial graph databases lack native temporal support, hindering their usefulness in these use cases.

This paper introduces Clock-G, a temporal graph management system designed to handle the history temporal graphs. What differentiates
Clock-G from other temporal graph management systems is its comprehensive approach, covering query language, query processing, and physical storage. We define T-Cypher, a temporal extension of Cypher query
language, enabling user-friendly and concise querying of the graph's history. Additionally, we propose a query processor that utilizes temporal statistics collected from underlying temporal graphs to offer a good
evaluation plan for T-Cypher queries. We also propose a novel storage
technique that balances space usage and query evaluation time.

Keywords: Temporal Graph management · Storage · Query
language · Query processing

1 Introduction

Graphs are frequently used to model real-world interactions as a collection
of vertices and relationships providing generally a fertile ground to analyze
relationship-centered domains. Despite the wealth of studies on managing static
graphs, a time version support is seldom provided.

This work is motivated by the industrial use case of Thing'in[1], an Orange-
initiated platform that manages a graph of connected (machines, traffic lights,
cameras, etc.) and non-connected (doors, roads, shelves, etc.) objects with structural and semantic environment descriptions. Clients include companies and

---

[1] https://www.thinginthefuture.com/.

© Springer-Verlag GmbH Germany, part of Springer Nature 2023
A. Hameurlain et al. (Eds.): *TLDKS LIV*, LNCS 14160, pp. 1–40, 2023.
https://doi.org/10.1007/978-3-662-68014-8_1

public administrations developing smart city services and private object owners building analytical IoT applications. The graph is maintained by a commercial database lacking temporal support. However, there has been an extensive and recent demand by the clients of Thign'in for preserving the past states and connections of the graph for the interest of tracking objects, anomaly detection and forecasting the future behaviour. Concrete use cases include Mo.Di.Flu[2], a project tracking product positions in a manufacturing pipeline to detect delays or losses. To address these requirements, we designed the temporal graph management system Clock-G. Although initially designed for the particular use case of Thing'in, Clock-G is a general purpose system that can be used in other application domains requiring temporal graph management.

Storing and querying temporal graphs are possible by exploiting a commercial graph database with temporal metadata [5,7]. However, these systems do not natively offer time-version support which might lead to unpredictable performances. Hence, we argue that time should be considered as a first-class citizen rather than a simple add-on.

Existing temporal graph management systems often lack comprehensive coverage of the different layers that should be addressed to account for the temporal dimension, as they may not provide a native temporal query language or an efficient query processor for temporal queries. Many existing systems [24,32,46] prioritize storage techniques and only offer simple, general-purpose temporal graph queries that cannot meet the requirements of specific applications such as the Thing'in use case. To address this issue, our paper takes a comprehensive approach to managing temporal graphs by addressing the different layers of query language, query processing, and physical storage.

This paper is an extension of our previous work [31]. A major improvement of this version compared to our previous work is the inclusion of a query language that supports complex temporal queries into Clock-G, including graph pattern matching and navigational queries with temporal predicates. Additionally, we have developed a query processor capable of evaluating temporal graph queries. Unlike the previous version, which focused primarily on storage techniques, this version addresses the challenges of query languages and processing, making our system more comprehensive.

Various temporal graph querying solutions have been proposed in the literature, extending OLAP and OLTP queries with time. OLAP queries include finding most durable connected components [42], temporal shortest paths [21], and temporal centrality [35], while OLTP queries include temporal graph pattern matching [33,40] and temporal navigational queries [2,37]. In this paper, we focus on extending OLTP queries with the temporal dimension. Hence, we propose T-Cypher, a temporal extension of the well-known graph query language Cypher [10] designed to enhance graph pattern matching and navigational queries with the temporal dimension.

Example 1.1 Figure 1 illustrates a graph pattern and its corresponding Cypher query, as well as a temporal graph pattern and its corresponding T-Cypher query.

---

[2] https://www.pole-emc2.fr/projet/mo-di-flu/.

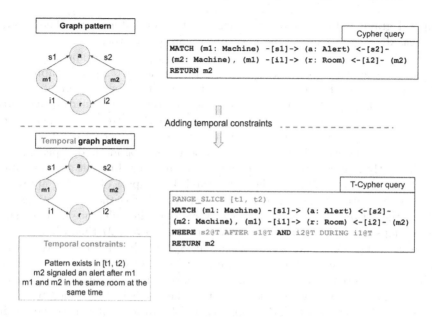

Fig. 1. Example showing a T-Cypher query with temporal constructs compared to a non-temporal Cypher query

In this example, the non-temporal pattern retrieves machines ($m1$ and $m2$) that indicate the same alert (a) and are situated in the same room (r). To enhance machine maintenance efficacy, one might want to identify machines that are affected by malfunctions in other machines. In such a scenario, the order in which alerts were triggered becomes significant since it allows for the retrieval of machines that indicated an alert after another machine signaling the same alert, thus implying machine-to-machine influence. This is translated in the T-Cypher query by the inclusion of the temporal constraints (`s2@T AFTER s1@T AND i2@T DURING i1@T`). Besides we trim the search space to a time interval [t1, t2) of interest instead of searching the full history which is translated in the T-Cypher query by `RANGE_SLICE` [t_1, t_2) clause at the beginning of the query.

We propose a query processor to evaluate T-Cypher queries. Our processing pipeline involves an algebra, cost model, and plan selection algorithm. We introduce a temporal graph algebra that extends the graph algebra proposed by Hölsch et al. [22] for Cypher queries with temporal operators. The evaluation plan composed of algebraic operators is chosen based on a cost model that relies on changing cardinalities provided by the backend store. Unlike traditional query processors, we consider the optimal plan to vary within the requested time interval due to changes in cardinalities, and thus preserve the history of cardinalities in temporal histograms. We implemented this query processor in Clock-G and evaluated it by executing various queries on synthetic datasets, comparing its performance with an alternative solution that is based on extending a non-temporal graph database (Neo4j[3]) with the temporal dimension. The results demonstrate the efficiency of our cost model and query processor.

Besides proposing a query language and a query processing pipeline, we propose and implement into Clock-G a storage technique for temporal graphs. Various storage approaches have been proposed in literature for managing temporal graphs, including the *Log* and the *Copy+Log* methods. The former involves preserving all graph updates as timestamped logs, while the latter stores the updates in time windows, along with snapshots (i.e. states of the graph) at the end of each window. However, the space usage of the *Copy+Log* method, especially for growth-mostly graphs, can be space-consuming due to redundant graph entities shared between snapshots. On the other hand, the impact of the *Log* approach on query evaluation time can be detrimental. To address these limitations, we propose the *δ-Copy+Log* method, which stores only the difference between successive snapshots, called deltas. Snapshots are stored every M time windows and used as starting points for query evaluation. Specifically, half of the time windows and their corresponding deltas are stored in a forward fashion, while the other half are stored in a backward fashion. During query evaluation, the choice between forward or backward construction of the result is determined based on the requested time instant. This approach results in a significant reduction in the maximum execution time of queries by up to 50%. We also conducted experiments to evaluate the performance of Clock-G. A comparison between traditional methods and the *δ-Copy+Log* validates that our technique offers a good compromise between the performances of the *Log* and *Copy+Log* methods. Besides, we showcase how the parameters of Clock-G can be calibrated in order to tune the overall performance with an adequate configuration that adheres most with the acceptable threshold of query latency and available storage resources.

The main contributions of this work reduce to the following:

- Proposing a user-friendly extension of the Cypher query language that enables to express a large fraction of temporal queries.
- Proposing a temporal graph algebra and query processor for T-Cypher queries.
- Proposing *δ-Copy+Log* as a space-efficient variant of the traditional *Copy+Log* method.
- Taking a holistic approach into managing temporal graphs by addressing the different layers of storage, query language, and processing.

Outline. Section 2 provides an overview of related work. Section 3 introduces key definitions for our proposed approaches. Section 4 introduces our proposed temporal graph query language. Section 5 describes the query processor used to evaluate our temporal graph queries. Section 6 presents our proposed storage approach. Section 7 presents the architecture and overall design features of Clock-G. Section 8 presents the results of our experiments conducted on real and synthetic datasets. Finally, Sect. 9 concludes the paper and gives a future perspective.

---

[3] https://neo4j.com.

2   Related Work

This paper proposes a comprehensive approach to managing temporal graphs with versioning support, which includes addressing challenges related to storage techniques, query languages, and query evaluation. We discuss the related work on these challenges in subsequent sections.

2.1   Query Language

In the field of graph querying, subgraph pattern matching or navigational queries are the core concepts. Many proposals to extend these queries with the temporal dimension have been posited.

Some extensions focus only on extending navigational queries with the temporal dimension. Temporal reachability queries where extended with the temporal dimension [41,45]. Granite [37] is a query engine that implements temporal navigational queries by adding temporal predicates and temporal ordering constraints, as well as temporal aggregations. A temporal extension of regular path queries (TRPQ) was proposed in [2] by introducing structural and temporal navigational operators. The T-GQL [7] query language is a temporal extension of the standard query language for graph databases GQL [8]. The proposed extension allows the expression of different types of temporal paths. However, these solutions focus on navigational queries rather than graph pattern matching queries.

Other proposals present temporal extensions of graph pattern matching queries. For instance, non-decreasing time flow pattern are defined as each path between two nodes follows a non-decreasing time flow [35,38,47]. It is useful for studying the spread of a disease or the flow of rumors in a social network. Most Durable Graph Pattern (MDGP) returns the most durable matches of a given non-temporal pattern, which is useful for analyzing the tightness of connectivity between nodes [42]. Despite the usefulness of these proposals in some applications, they cannot express more general temporal predicates between the elements of a pattern.

The Temporal Graph Algebra (TGA) [33] is a temporal generalization based on temporal relational algebra for some graph operators. These operators can filter the search to a time instant or interval or returns subgraphs that are isomorphic to a given pattern during a given time instant or interval. GRALA [40] is a temporal analytical language that offers temporal operators to determine graph snapshots, the difference between two snapshots, and the subgraphs satisfying a given time-dependent graph pattern. Despite the novelty of these extensions, they do not support navigational queries.

T-SPARQL [13] is a temporal graph query language for RDF that embeds the features of TSQL2 [43]. To express temporal predicates over timestamp variables, the authors propose using a subset of Allen's temporal operators [1]. SPARQL^T is another query language for temporal RDF stores. Although this language does not offer dedicated temporal operators to express temporal relations, it is possible to use temporal functions that extract the starting and ending time instants of a tuple to express any of Allen's temporal relations and temporal slicing. In this paper, temporal predicates and slicing are used on the property graph model, extended to include temporal navigation functionalities.

This paper presents a novel method for querying temporal graphs that builds upon graph pattern matching and navigational queries by incorporating the temporal dimension. Our primary objective is to propose a concise and user-friendly syntax that is easy to learn and facilitates intuitive reasoning and query construction for temporal graphs. Motivated by this goal, we proposed T-Cypher (Sect. 4), a temporal graph query language that extends the popular Cypher query language [10]. The rationale behind this choice is that the syntax of Cypher is graph-like (i.e., graph patterns are expressed using "ASCII art") and user-friendly, making it a popular choice amongst graph query languages. Many features extracted from Cypher will be echoed in the standardization of upcoming standrad graph query language GQL [8]. Besides, Cypher is expressive, declarative, normalized, and open source.

2.2 Query Processing

A query processor uses an algebra to convert a query into a set of algebraic operators. In the context of temporal graph management, a Temporal Graph Algebra (TGA) was proposed in [33], which includes graph operators that are extended with the temporal dimension. In this work, we define a temporal graph algebra that extends the graph algebra defined by Hölsch et al. for Cypher queries in [22]. Our choice of extending this algebra, rather than other alternative graph algebras (such as GraphQL [19] and GRAD [12]), is based on its compatibility with our proposed query language, which extends Cypher.

Evaluating a query implies choosing a good evaluation plan that ideally minimizes the cardinality of sub-results, reducing thus the overall execution time. The plan selection technique is usually coupled with a cost model that defines a cost function for each algebraic operator which allows to approximate the resulting cardinality of an operator before evaluation. This evaluation pipeline was followed in [14] for processing Cypher queries in a graph database. However, our goal in this paper is to extend this pipeline for the temporal graph model.

A query processor for temporal navigational queries can be found in Granite [37]. This plan selection approach splits the query path into sub-path segments to reduce cardinality, and uses a cost model based on temporal histograms to estimate plan cost. However, this approach is limited to path queries and cannot handle temporal graph pattern matching, which requires a more complex plan selection approach. To address this problem, we present a query processor that evaluates temporal graph pattern matching queries (Sect. 5).

2.3 Storage

Available temporal graph storage techniques can be categorized as follows: **Log**, **Copy**, **Copy-On-Write** and **Copy+Log**. These methods are mainly motivated by concepts of logging and *checkpointing* which reflects on lessons learned from classical techniques of database state recovery. The **Log** storage approach used in [11,16] stores graph updates as timestamped logs, allowing recovery of any graph state by loading logs with a timestamp lower than or equal to the requested one. In contrast, the **Copy** approach materializes and persists graph snapshots. These methods represent two extremes in storing temporal graphs, favoring either space optimization or query computation time optimization. **Copy-On-Write** [4,20,28,30] involves copying a single graph entity whenever it gets updated, while **Copy+Log** [17,18,24–26,32,46] stores graph updates in temporally disjoint partitions (called time windows), along with snapshots representing valid states of the graph. The advantage of the Copy+Log approach is that the state of the graph at a given time instant can be recovered by reading a single snapshot and all graph updates recorded in a time window.

In this work, we address a critical limitation of the *Copy+Log* storage approach which relates to the high space consumption of full graph snapshots. To mitigate this issue, we propose the δ-**Copy+Log** (Sect. 6) approach which considers the difference between snapshots instead of materializing full snapshots. We preserve a number of snapshots to serve as starting points for query evaluation and after a fixed number of delta, a full snapshot is materialized. This approach differs from traditional methods such as RMAN in that it replaces full backups with deltas that contain only the difference between two snapshots.

3   Formal Definitions

3.1   Time Domain

In this section, we present a definition of the time domain, which is essential in the development of data management systems that incorporate temporal ontologies. The time domain definition is particularly important in assigning temporal validity information to data items [34]. Our approach to modeling time involves selecting a discrete temporal flow, which is achieved by quantifying a time axis with time granules [6]. Time granules, also known as chronons, are the smallest indivisible units of time defined by a specific temporal granularity (such as a second or a millisecond). We define the time domain, denoted as Ω^T, as a totally ordered set of instants that includes a sequence of discrete time granules: $\Omega^T = \{t^i | i \in \mathbb{N}\} \bigcup \{Now, \infty\}$. The duration between consecutive instants in the sequence is equal to a chronon. In addition, we assume that the system assigns a transactional time to each graph update.

3.2   Temporal Graph Relation

A T-Cypher query produces a temporal graph relation as output. Each relation is represented as a bag of tuples, where a tuple u is a partial function that maps names to values. The named fields of a tuple are defined as $u = (a_1 : v_1, \ldots, a_n : v_n)$, where $(a_1, \ldots, a_n)$ are distinct names, and each element in $(v_1, \ldots, v_n)$ can be a value, node or relationship state, set of node or relationship states, or paths.

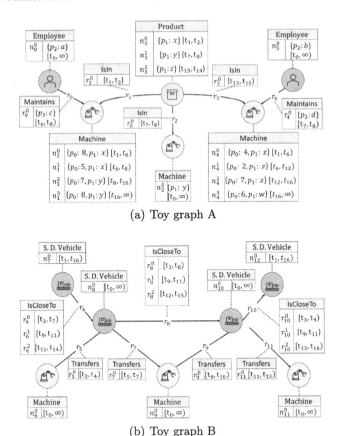

(a) Toy graph A

(b) Toy graph B

Fig. 2. Toy graphs illustrating the traversal of products through machines, the maintenance of these machines (Toy graph A), the transfer of products between machines, and the closeness between self driving vehicles (Toy graph B)

These states correspond to nodes or relationships within a specific time interval during which their property values remained constant. We consider V, k, ID, L, and T to denote the set of values, property keys, node identifiers, node labels, and relationship types, respectively.

A **node state** in n is a tuple (id_n, l, k, τ) such that:

- $id_n \in ID$ is the node identifier.
- $l \in 2^L$ is the set of node labels.
- $k = \{k_1 : v_1, \ldots, k_m : v_m\}$ is a map of property names and values such that $k_i \in k$ and $v_i \in V, \forall 1 \le i \le m$.
- $\tau \in \Omega^T \times \Omega^T$ is the validity time interval during which the node state was valid.

A **relationship state** in r is a tuple $(id_{n_s}, id_{n_t}, t, k, \tau)$ such that:

- $id_{n_s} \in ID$ is the source node identifier.

- $id_{n_t} \in ID$ is the target node identifier.
- $t \in 2^T$ is the set of relationship types.
- $k = \{k_1 : v_1, \ldots, k_m : v_m\}$ is a map of property names and values such that $k_i \in k$ and $v_i \in V, \forall 1 \le i \le m$.
- $\tau \in \Omega^T \times \Omega^T$ is the validity time interval during which the relationship state was valid.

Example 3.1 To clarify the previous definitions, we present a concrete example of a temporal property graph inspired by the use case of smart factories. Figures 2(a) and 2(b) show two toy graphs (A and B) inspired by this use-case. In these graphs, nodes $\{n_0, \ldots, n_{12}\}$ model products, machines, self-driving vehicles (S.D. vehicles), and employees. Whereas, relationships $\{r_0, \ldots, r_{11}\}$ represent the connections between nodes. Properties $\{p_0, \ldots, p_3\}$ are attached to nodes and relationships to describe these graph entities.

In the manufacturing process, a product may traverse through various machines, which is denoted by the *isIn* relationship between them. This relationship captures the progression of the product through the different stages of the manufacturing process. The machines are regularly maintained by employees through the *maintains* relationship. Additionally, products are transported from one machine to another through an S.D. vehicle using the *transfers* relationship. To indicate the proximity between S.D. vehicles, a temporary relationship *isCloseTo* is established if the distance between them is lower than a predetermined threshold. The properties p_0 and p_1 of a machine can indicate its temperature or position whereas the property p_2 of an employee can its skills. The tools used during maintenance can be represented by p_3 of the maintains relationship. Each node and relationship in the temporal graph contains several states that map property names to values during specific time intervals. Querying this temporal graph allows for analyzing the causes of system failures by tracking the trajectory of products and monitoring the evolution of machine states. We present in Tables 1(a) and 1(b) the node and relationship states of $N = \{n_0, \ldots, n_{12}\}$ and $R = \{r_0, \ldots, r_{11}\}$ of the temporal property graphs (A and B) presented in Fig. 2.

Let us now discuss the creation of node states $\{n_1^0, n_1^1, n_1^2, n_1^3\}$ in the Toy graph A (Fig. 2(a)). For instance, the first node state n_1^0 is bound with values $(8, x)$ for property keys (p_0, p_1). This state is valid during $[t_1, t_6)$ since an update of the property p_1 occurred at time instant t_6 which results in a new node state n_1^1. Both node states have the same value for the unmodified property (p_1) and different values for the updated property (p_0). Similarly, the node state n_1^2 is created after the update of the properties p_0 and p_1 at time instants t_8. Finally, the last modification of the node is an update of the property p_0 at time instant t_{16} which results in a new node state n_1^3 valid in $[t_{16}, \infty)$.

Table 1. A fraction of the relationships and their states of the graph in Figs. 2(a) and 2(b)

(a) Node states

Nodes	States
n_0	$n_0^0 = (id_{n_0}, \text{Employee}, \{p_2 : a\}, [t_0, \infty))$
n_1	$n_1^0 = (id_{n_1}, \text{Machine}, \{p_0 : 8, p_1 : x\}, [t_1, t_6))$
	$n_1^1 = (id_{n_1}, \text{Machine}, \{p_0 : 5, p_1 : x\}, [t_6, t_8))$
	$n_1^2 = (id_{n_1}, \text{Machine}, \{p_0 : 7, p_1 : y\}, [t_8, t_{16}))$
	$n_1^3 = (id_{n_1}, \text{Machine}, \{p_0 : 8, p_1 : y\}, [t_{16}, \infty))$
n_2	$n_2^0 = (id_{n_2}, \text{Machine}, \{p_1 : y\}, [t_0, \infty))$

(b) Relationship states

Relationships	States
r_0	$r_0^0 = (id_{n_0}, id_{n_1}, \text{Maintains}, \{p_3 : c\}, [t_6, t_8))$
r_1	$r_1^0 = (id_{n_3}, id_{n_1}, \text{IsIn}, \{\}, [t_1, t_2))$
r_2	$r_2^0 = (id_{n_3}, id_{n_2}, \text{IsIn}, \{\}, [t_7, t_8))$

4   Temporal Graph Query Language

This section presents our temporal graph query language T-Cypher that extends Cypher with temporal constructs. Throughout this section, we provide clear query examples and their results to clarify the semantics of our temporal constructs. A more detailed description of the syntax and semantics of T-Cypher is given the online documentation[4]. Our proposed extension, T-Cypher, is designed to incorporate temporal constructs without requiring modifications to the existing grammar rules. This approach ensures a straightforward transition for practitioners who are already familiar with Cypher, while reducing query verbosity.

With T-Cypher, graph variables such as nodes, relationships, and properties, as well as temporal variables referring to time validity intervals, can be expressed. This enables the application of temporal constraints to the temporal variables of the query. Furthermore, T-Cypher introduces the trim statement, which can be used at the beginning of a query to prune the search space to single or multiple time intervals. This guarantees that all variables defined in the query are valid during at least one of these intervals. Temporal functions and operators, can also be used in T-Cypher to define constraints and predicates on temporal variables of the query. Another key feature of T-Cypher is the ability to express different types of temporal paths. We define these key temporal constructs in the following.

Temporal Slicing Clause. We propose a temporal slicing clause to prune the search space of a query to a single time instant or time interval. Hence, the tem-

---

[4] https://project.inria.fr/tcypher/.

poral selection will be applied to all the variables of a temporal query such that the returned states of graph entities should be valid at the requested time instant or during the requested time interval. We use different time slicing techniques using the tokens SNAPSHOT, RANGE_SLICE, LEFT_SLICE and RIGHT_SLICE.

A query starting with the SNAPSHOT token searches for graph entities that are valid at a single requested time instant. On the other hand, a query starting with a time-slicing token, such as RANGE_SLICE, LEFT_SLICE, or RIGHT_SLICE, searches for graph entities whose time intervals intersect with the requested time interval, starts before, or ends after the requested time instant, respectively. If a query does not start with a time-slicing token, it is applied to the latest version of the graph.

Figure 3 shows two queries applied to the Toy graph A (Fig. 2(a)), with their results. The first query returns the machine states valid at t_3, while the second query returns machine states with time intervals intersects with $[t_1, t_8)$.

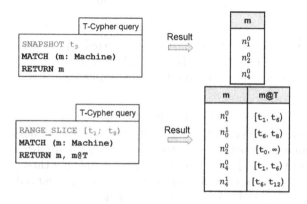

Fig. 3. Example of temporal slicing

Temporal Functions and Operators. We define a set of temporal functions that can be applied to the temporal variables of a pattern to define temporal predicates. For space limitations, we present some of these functions in Table 2, whereas a more comprehensive description is given in the online documentation of T-Cypher. Besides, we use Allen's operators [1] (e.g., before, after, during) to define temporal relations between the temporal variables of a pattern.

Figure 4 provides an example of a T-Cypher query using temporal functions and operators and its result when applied to the Toy graph A (Fig. 2(a)). This query returns the elapsed time[5] between the maintenance of a machine and its failure. The failure of a machine can be detected if the value of property p_0 (e.g., temperature) is higher than a threshold. The expression (n@T AFTER e@T)

---

[5] The elapsed time between two time intervals i and i' is equal to the difference between the starting time instant of i' and the ending time instant of i.

Table 2. Description of some temporal functions used in T-Cypher

Function	Description	Return type
ELAPSED_TIME(i, i')	Returns the elapsed time between i and i'	Duration
DURATION(i)	Returns the duration of i	Duration
INTERSECTION$(i_0, \ldots, i_n)$	Returns the intersection between $\{i_0, \ldots, i_n\}$	interval

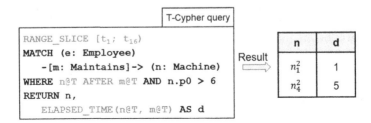

Fig. 4. Example of temporal functions and operators

indicates that the system failure must have occurred after the maintenance. We notice that the machine state n_1^2 is returned since it has a value of p_0 higher than the threshold and it occurred after the maintenance of the machine.

Temporal Paths. The relationships in a temporal graph are valid during certain time intervals. Hence, the connectivity between two nodes can be subject to temporal conditions defined over the relationships of a path which results in diverse types of temporal paths. In T-Cypher, we include three temporal types that can cover a large subset of queries: Continuous, Sequential, and Pairwise-continuous (Fig. 5), which we describe in the following.

Temporal Path. A temporal path is defined as a tuple $(n_1^s, r_1^s, \ldots, r_k^s, n_{k+1}^s, \tau_p)$ containing a sequence of k relationship states $(r_i^s, \forall 1 < i < k)$ and $k + 1$ node states $(n_i^s, \forall 1 < i < k+1)$ and a time interval during which the path is valid. Each relationship state $(r_i^s, \forall 1 < i < k)$ is a tuple $(id_{n_i}, id_{n_{i+1}}, t_{r_i^s}, k_{r_i^s}, \tau_{r_i^s})$ connecting two node states of the path $n_i^s = (id_{n_i}, l_{n_i^s}, k_{n_i^s}, \tau_{n_i^s})$ and $n_{i+1}^s = (id_{n_{i+1}}, l_{n_{i+1}^s}, k_{n_{i+1}^s}, \tau_{n_{i+1}^s})$. The time interval of the path τ_p is derived from the time intervals of the path relationships and depends on the type of the temporal path.

Continuous Path [39]. A continuous path is a temporal path where the intersection between the time intervals $(\tau_{r_i^k}, \forall 1 < i < k)$ of the relationship states (r_i^k) of the path is not null and τ_p is equal to the intersection between time intervals $\{\tau_{r_1^s}, \ldots, \tau_{r_k^k}\}$.

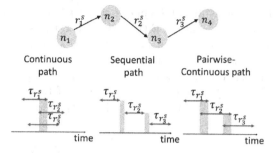

Fig. 5. Different types of temporal relationship patterns: Continuous, Pairwise Continuous and Sequential ($\tau_{r_1^s}$, $\tau_{r_2^s}$ and $\tau_{r_3^s}$ refer to time validity intervals of relationships r_1^s, r_2 and r_3^s)

Figure 6 presents a T-Cypher query with a continuous path and its result when applied to Toy graph B (Fig. 2(b)). This query returns the path between self-driving vehicles that were 3-Hop close to each other during the time interval $[t_1, t_{16})$. Hence, the self-driving vehicles of the path were close during the intersection of the time intervals of the path relationships. Notice that three continuous paths of length 3 exist between the self-driving vehicles n_7 and n_{12}. The time interval $[t_3, t_7)$ of the first path is equal to the intersection between the time intervals of its relationship states ($[t_3, t_7)$, $[t_3, t_8)$ and $[t_3, t_9)$).

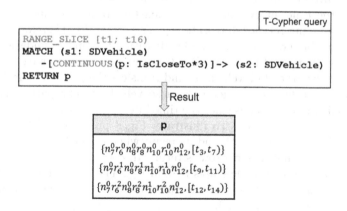

Fig. 6. Example of a continuous path

Sequential Path [23,38,47]. A sequential path is a temporal path where each relationship state r_{i+1}^s should occur after the relationship state r_i^s ($\forall\, 1 \leq i < k$). Hence, the ending time instant of $\tau_{r_i^s}$ should be lower than the starting time instant of $\tau_{r_{i+1}^s}$. The time interval of the path is the range of time covered by the time intervals of the path.

To illustrate, Fig. 7 presents a T-Cypher query with a sequential path and its result when applied to the Toy graph B (Fig. 2(b)). It returns a product's transfer path of length 4 between two machines, implying that a self-driving vehicle or a machine transfers a product after receiving it. Note that a sequential path of length 4 exists between the node states n_3 and n_9. This path is valid during the time interval $[t_2, t_{13})$ that represents the range of the time intervals of its relationship states $([t_2, t_4), [t_5, t_7), [t_8, t_{10})$ and $[t_{11}, t_{13}))$.

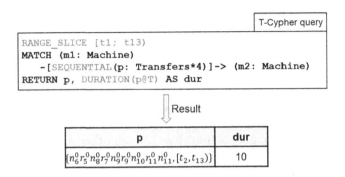

Fig. 7. Example of a sequential path

Pairwise-Continuous Path [7]. A pairwise-continuous path is a temporal path where the time interval of each relationship state r_i^s should overlap with that of the outgoing relationship state r_{i+1}^s ($\forall\, 1 \leq i < k$). Therefore, $\tau_{r_i^s}$ starts within the time boundaries of $\tau_{r_{i-1}^s}$ and ends within the time boundaries of $\tau_{r_{i+1}^s}$.

Let us now consider that a vehicle a transfers a product to a close vehicle b. Now, b also looks for a close vehicle, c, and transfers the product to it. Similarly, the vehicle c transfers a product to a close vehicle d. The path between the vehicles is pairwise continuous since the time intervals of each pair of consecutive relationships are overlapping. To illustrate, Fig. 8 presents a T-Cypher query with a pairwise-continuous path and its result when applied to the Toy graph B (Fig. 2(b)). Notice that a single row is returned, corresponding to the pairwise-continuous path between node states n_7 and n_{12}. The time interval of the path $[t_{11}, t_{16})$ is equal to the range of the time intervals of its relationship states $([t_{11}, t_{14}), [t_{12}, t_{15})$ and $[t_{13}, t_{16}))$.

5   Temporal Graph Query Processor

In this section, we give an overview of the query processing pipeline presented in Fig. 9.

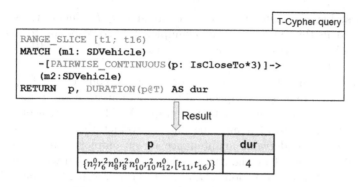

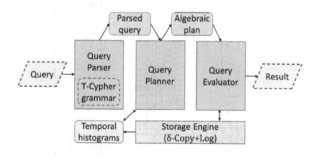

Fig. 8. Example of pairwise-continuous path

The query parser checks the syntax of a T-Cypher query according to defined grammar rules and generates an Abstract Syntax Tree (AST). The parser then uses the AST to create a parsed query object that is understandable by the query planner. Using a cost-based model, the query planner generates an algebraic plan with cardinalities of subqueries based on temporal histograms. The query evaluator executes each query operator by communicating with the storage engine using δ-Copy+Log technique presented in Sect. 6. In the following, we will introduce our temporal graph algebra, cost model, and plan selection algorithm.

Fig. 9. The query processing pipeline implemented in Clock-G

5.1 Temporal Graph Algebra

We extend the graph algebra proposed by Hölsch et al. [22] by adding time-based operators to translate T-Cypher queries into algebraic representations. Our extension relies on temporal graph relations defined in Sect. 3.2. These relations are bags of tuples that map names to various entities, including node or relationship states, sets of states, or temporal paths.

Operators. Let E denote an algebraic expression, $\mu(E)$ denote the set of variables defined in the expression. For example, if E corresponds to matching a relationship between two node variables (a and c) such as ($a - [b]- > c$), then $\mu(E)$ is the set of variables $\{a, b, c\}$.

We illustrate the utilization of our operators to convert a T-Cypher query into an algebraic expression. Specifically, we demonstrate the process using a sample query Q that is applied to toy graph A. The objective of this query is to retrieve the state of a machine and a product that was present in the machine before it underwent maintenance by an employee during a specified time period. We present in Fig. 10 a possible evaluation plan with the results of the different algebraic expressions ($\{E_0, \ldots, E_5\}$) composing the plan. Note that these expressions are given in the following description of the operators.

```
Query Q

RANGE_SLICE [t1; t8]
MATCH (m: Machine) <-[r: Maintains]- (e: Employee),
(m) <- [i: IsIn] - (p:Product)
WHERE m.p_0 > 2 AND m@T BEFORE r@T
AND i@T BEFORE r@T AND p@T DURING i@T
RETURN m, r, e, i, p;
```

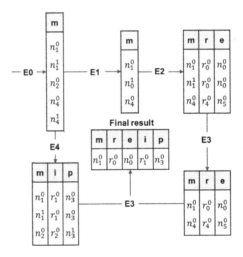

Fig. 10. Example showing the result of evaluating the algebraic operators ($E_0, \ldots, E_5$) on the toy graph in Fig. 2(a)

GetNodes Operator. The GetNodes operator returns a temporal graph relation containing node states from the underlying graph G. We use $\bigcirc_{\tau,a,\rho_a}$ to denote this operator, where

- τ: is a time interval such that the returned node states should have time intervals that overlap with it.
- a: is the name of the node variable.
- ρ_a: is the label of the node variable.

Every node state in the underlying graph having a label ρ_a and its time interval overlaps with τ will be returned by this operator. To illustrate this operator, we consider the following expression returning the machine states valid in $[t_1, t_8)$.

$$E_0 = \bigcirc_{[t_1,t_8),m,Machine}$$

The result of this operator is given in Fig. 10.

Select Operator. The select operator, denoted as $\sigma_{\tau,\theta}(E)$, filters input tables based on property values of node or relationship states. It uses a Boolean expression θ defined over validity intervals and property values of variables from $\mu(E)$. This operator filters tuples from input graph relations that satisfy θ during the time interval τ. To illustrate this operator, consider the following expression:

$$E_1 = \sigma_{[t_1,t_8),m.p_0>2}(E_0)$$

This operator filters the input relation resulting from applying E_1 such that the value of the property p_0 of m is lower than 2. The result of this operator is illustrated in Fig. 10.

Expand Operator. The expand operator creates a new relation by expanding input relation tuples with direct relationships and target nodes. It is denoted as $\uparrow_{\tau,a,b,ab,\rho_b,\rho_{ab}}(E)$, and ensures that added relationship states are valid within a specified time interval, where

- τ: is a time interval such that the returned relationship states should have time intervals that overlap with it.
- a: is the name of the node in the input relation.
- b: is the name of the added target node.
- ab: is the name of the added relationship.
- ρ_b: is the label of the added target node b.
- ρ_{ab}: is the type of the added relationship ab.

To denote an expansion with an incoming direction, we write $\downarrow_{\tau,a,b,ab,\rho_b,\rho_{ab}}(E)$.

The expand operator can express joins between the input relation and underlying graph when a node in the input relation reaches another node in the graph through a relationship. However, expressing paths through a recursion of join operators leads to a limited relational model. The expansion operator is more general and convenient, as it does not restrict the data model. To illustrate this operator, consider the following operator:

$$E_2 = \downarrow_{[t_1,t_8),m,e,r,Employee,Maintains}(E_1)$$

Consider that this operator's input is the previous expression E_1 resulting from the select operator. Notice that the node states (n_1^0 and n_1^1) are each expanded with (r_0^0 and n_0^0) and the node state n_4^0 is expanded with (r_4^0 and n_5^0). Let us filter the returned result to keep the machine states valid before the maintenance, as follows:

$$E_3 = \sigma_{[t_1,t_8),\text{m@T BEFORE r@T}}(E_2)$$

The result of this operator is given in Fig. 10.

Join Operator. The Join operator joins two expressions based on a Boolean expression. We use $E \bowtie_\theta E'$ to denote this operator where θ is a Boolean expression. To illustrate this operator, consider joining the previously described expression E_3 with the expression E_4 given below. This expression returns the product states valid when the product was in a machine in $[t_1, t_8)$.

$$E_4 = \sigma_{[t_1,t_8),\text{p@T DURING i@T}}(\downarrow_{[t_1,t_8),m,i,p}(\bigcirc_{[t_1,t_8),m}))$$

The following operator joins E_3 and E_4 with a temporal condition. We refer to a junction with a temporal condition as a temporal join. The result of this operator is illustrated in Fig. 10.

$$E_5 = E_3 \bowtie_{\text{i@T BEFORE r@T}} E_4$$

It should be mentioned that more complex operators can be defined including the aggregation operator that we keep for later work.

5.2   Cost Model

This section defines the cost model used by the query planner, which estimates the cost of an evaluation plan. The cost of each operator is equal to the estimated cardinality of its output relation. Our model differs from classical query processing models commonly used in relational databases because it considers the cost of a query to change over time, meaning an optimal plan for one time interval may not be optimal for another due to changing cardinalities. Our query planner accounts for this by computing the cardinality of each algebraic operator based on the requested time interval. We use $card(E)$ to denote the estimated cardinality of an algebraic expression E.

We use temporal histograms to estimate the cardinalities of algebraic operators for a given requested time. We create a temporal histogram for the **evolution** of each of the following:

- Number of node states with a given label.
- Number of relationship states with a given type and labels for the source and target nodes.
- Number of node states with a given label and a value for a property name.
- Number of relationship states with a given type and a value for a property name.

GetNodes Operator. The cost of the getNodes operator is equal to the estimated cardinality of the node states with a given label ρ_a valid during a given time interval τ, as given in the equation below.

$$card\left(\bigcirc_{\tau,a,\rho_a}\right) = C_{(\tau,\rho_a)}$$

Expand Operator. The cost of the expand operator is equal to the average cardinality of the relationship states given the label of the source and target node states (ρ_a, ρ_b), type of the relationship state (ρ_{ab}), requested time interval (τ) multiplied by the cost of the previous expression E (card(E)), as given in the equation below.

$$card\left(\uparrow_{\tau,a,b,ab,\rho_b,\rho_{ab}}(E)\right) = \frac{C_{(\tau,\rho_a,\rho_b,\rho_{ab})}}{C_{(\tau,\rho_a)}} * card(E)$$

Select Operator. The cardinality of the select operator applied on an expression E is equal to the selectivity of the graph entity states $sel_\theta(E)$ satisfying the given condition θ multiplied by the cardinality of E, as given in the equation below.

$$card(\sigma_{\tau,a.p=v}(E)) = sel(\tau, \rho_a, p, v) * card(E)$$

The selectivity of graph entities is computed as follows:

$$sel(\tau, \rho_a, p, v) = \frac{C_{(\tau,\rho_a,p,v)}}{C_{(\tau,\rho_a)}}$$

Note that, we define the cost of the selection operator in which we only consider filtering on the values of the node and relationship properties defined in the input expression. The selectivity of a condition θ is equal to the cardinality of all the graph entities satisfying it $a.p = v$ divided by the cardinality of all graph entities with a label or type ρ_a that existed during the time interval τ.

Join Operator. The cost of the join operator applied on expressions E, and E' is equal to the product of the cardinalities of these expressions, as given in the equation below.

$$card(E \bowtie E') = card(E) * card(E')$$

5.3   Greedy Plan Selection Algorithm

This section describes an algorithm that greedily generates an evaluation plan for a T-Cypher query (Algorithm 1). The main idea is to iteratively compute the optimal plan such that an optimal decision is chosen at each iteration by selecting the less costly algebraic operator and adding it to the final plan.

The input is a query object Q, whereas the output is the algebraic plan p_{final}. The first step is to compute all the GetNodes operators representing the leaves of the logical plan tree and add them to the set of sub-plans P. In each iteration, a candidate set P_{cand} is initialized, which will then contain the possible operators

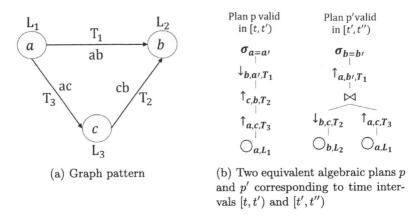

(a) Graph pattern

(b) Two equivalent algebraic plans p and p' corresponding to time intervals $[t, t')$ and $[t', t'')$

Fig. 11. Example illustrating a graph pattern and two possible logical plans, each corresponding to a time interval

that can be applied to the set of sub-plans P to cover all the nodes defined in the query. Then, the possibility of joining two sub-plans is first checked, and every possible join operator is added to the candidate plans P_{cand}. The method **joinExists**(p, p') returns the following:

$$joinExists(p, p') = \begin{cases} True, & \text{if } \mu(E_p) \cap \mu(E_{p'}) \neq \{\} \\ False, & \text{Otherwise} \end{cases}$$

Consider $\mu(E_x)$ to denote the set of variables of the expression of the plan x, then the method returns true if the variables of the plan p intersect with the set of variables of the plan p' and false otherwise. After including the possible joins in P_{cand}, each candidate plan is extended with an Expand operator such that the added node variable does not exist in the original plan. Every extended plan will be added to P_{cand}. Now, if no candidate operators are available, the final plan, which encloses all the node variables of the query Q is found, and the iterations stop. Otherwise, the most optimal plan p_{opt} is chosen between the set of candidate plans P_{cand} such that the cost of each plan corresponds to the requested time interval τ. Note that the computations of the costs of each operator are described in Sect. 5.2. After adding p_{opt} to P, the other plans contained in P and enclosed by p_{opt} are removed from P. The method **enclose** returns the following:

$$p.enclose(p') = \begin{cases} True, & \text{if } \mu(E_p) \supseteq \mu(E_{p'}) \neq \{\} \\ False, & \text{Otherwise} \end{cases}$$

This implies that a plan p is considered to enclose another plan p' if the set of variables of p contains all the variables of p'. Finally, the iterations stop when no candidate sub-plans are added to the P_{cand} and the final plan p_{final} contained in P is returned.

Algorithm 1: Greedy selection of a logical plan for a T-Cypher query

---

Input: Query object Q, τ
Output: Logical plan p_{final}

1   $P \leftarrow$ InitPlans() $N \leftarrow$ ExtractNodes(Q) ;
2   $\tau \leftarrow$ ExtractTimeInterval(Q) ;
3   **for** $n \leftarrow N$ **do**
4      $p \leftarrow$ getNodes(n) ;
5   P.insert(p) ;
6   Do$P_{cand}.size \geq 1$ $P_{cand} \leftarrow$ initPlans() ;
7   **for** $p \in P$ **do**
8   **for** $p' \in P$ **do**
9   **if** $joinExists(p, p')$ **then**
10            $p'' \leftarrow$ join(p, p') ;
11   P_{cand}.insert(p'') ;

12   **for** $p \in P$ **do**
13      $p' \leftarrow$ expand(p) ;
14   P_{cand}.insert(p') ;
15   **if** $P_{cand}.size \geq 1$ **then**
16      $p_{opt} \leftarrow$ chooseOptimal(P_{cand}, τ) ;
17   P.insert(p_{opt}) ;
18   **for** $p \in P$ **do**
19   **if** $p_{opt}.enclose(p)$ **then**
20   P.remove(p) ;

21   $p_{final} \leftarrow P$.get(0)

We show how an optimal plan for the graph pattern presented in Fig. 11(a) is computed by applying Algorithm 1. In this example, we present three node variables $\{a, b, c\}$ labelled with $\{L_0, L_1, L_2\}$ and the relationship variables $\{ab, cb, ac\}$ having types $\{T_0, T_1, T_2\}$. We show in Fig. 11(b) two of the many possible execution plans for this graph pattern. We assume that the cardinalities of the graph entities change over time which conduces to a change of the (greedily) optimal plan. Hence, we consider that the plans presented in Fig. 11(b) correspond to time intervals $[t, t')$ and $[t', t'')$, respectively. Figures 12(a) and 12(b) present the selection of operators in each iteration of Algorithm 1, yielding to plans p and p' presented in Fig. 11(b). Note that we omit some parameters from the notations of operators when they can be derived from the context.

Extracting cardinalities from temporal histograms implies fetching the total number of graph entity states with given constraints (node label or relationship type) that were valid during the requested time interval (i.e., their time intervals overlap with the requested time interval). A possible way of handling this is to keep all the cardinalities in an array such that querying it for a given time interval implies reading all the records until reaching the end time instant of the requested time interval. Despite its compact space usage, an array data structure

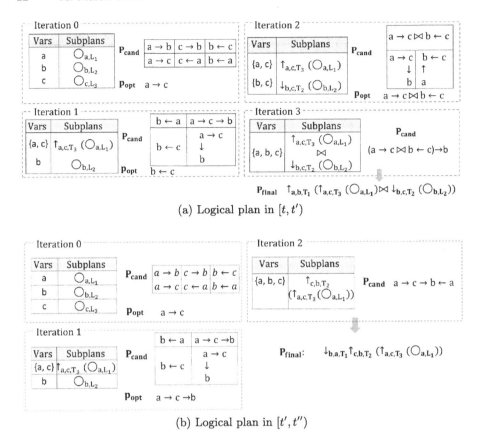

(a) Logical plan in $[t, t')$

(b) Logical plan in $[t', t'')$

Fig. 12. Greedy selection of logical plans in for different time intervals

implies at most searching all the elements of the array to retrieve the cardinality for a single time interval. To mitigate this complexity, we propose the use of segment trees [3].

A segment tree is a data structure that keeps information related to intervals as a full binary tree to allow an efficient response to range queries. For example, querying a segment tree allows finding an aggregated value (e.g., sum, maximum, average) of consecutive array elements in a range. For our query planner, we use segment trees to compute the maximum cardinality recorded in a time range to estimate the overall cost of a query plan. We choose the maximum cardinality since it can result in worst-case cost estimation.

6   Temporal Graph Storage

The *δ-Copy+Log* is a variant of the Copy+Log storage approach that we propose to mitigate the space cost induced by storing full snapshots. Recall that the Copy+Log consists of storing snapshots that are valid between the boundaries of a time window s.t. each time window contains a fixed number of graph operations. Now, the δ-$Copy$+Log follows a similar mechanism with the main difference that consists of storing deltas instead of snapshots. A critical point is that a delta differs from a time window. That is, a time window contains every graph operation that exists between two snapshots whereas a delta contains the only the minimum number of graph operations that transform a snapshot into another one. Indeed, an addition of an element in cancelled by a deletion of the same element, hence, both operations are stored in time window but omitted from the delta. We store a snapshot after a number of time windows in order to serve as a starting point for query evaluation. Having this, we store graph operations in consecutive time buckets containing each a number M of time windows such that the first $M-1$ time windows end with a delta, whereas the final time window ends with a snapshot. A critical optimization is the forward and backward data storage and retrieval. That is, half of the deltas and time windows in a bucket is constructed in a forward fashion whereas the other half is constructed in a backward fashion. The rationale behind this choice is the acceleration of the query execution time. That is, we choose the closest snapshot from which to start the retrieval then compute the result in a forward or backward fashion whether the time instant of that snapshot is lower or greater than the requested one.

Figure 13 illustrates the storage internals of the δ-$Copy$+Log and $Copy$+Log methods. It shows that the $Copy$+Log method stores time windows and snapshots. Whereas, the δ-$Copy$+Log stores time windows, deltas and snapshots. In this example, we consider a set of time buckets B where $M=6$ which implies that the bucket contains 3 forward time windows $\{\omega^1_\Rightarrow, \omega^2_\Rightarrow, \omega^3_\Rightarrow\}$ and 3 backward time windows $\{\omega^4_\Leftarrow, \omega^5_\Leftarrow, \omega^6_\Leftarrow\}$. At the highest time instant of a forward time window, a delta is materialized resulting in 2 forward deltas $\{\delta^1_\Rightarrow, \delta^2_\Rightarrow\}$. Whereas, a delta is materialized at the lowest time instant of every backward time window except the last time window where a snapshot is materialized resulting in 2 backwards deltas $\{\delta^4_\Leftarrow, \delta^5_\Leftarrow\}$ and snapshot $\{S^6\}$. Note that, the subtractive relation $\ominus$ operating on two snapshots S and S' s.t. $S \ominus S'$ results in the minimum number of graph updates that permits the transformation of S in to S'. Half of the time windows is stored in forward fashion whereas the othe'r half is stored in a backward fashion. Suppose a query with a requested time instant t. If t falls within the time interval of time window $\omega^2_\Rightarrow$, we start the search in a forward fashion by fetching $\omega^1_\Rightarrow$, then fetching $\omega^2_\Rightarrow$ whose timestamp is lower than t. Whereas, if t falls within the time interval of the time window $\omega^5_\Leftarrow$, we construct the result in a backward fashion. That is, we start by fetching S^6 then $\delta^5_\Leftarrow$ and finally $\omega^5_\Leftarrow$. Note that, in the $Copy$+Log method all the time windows are considered as forward.

In the following, we describe the key components of the δ-$Copy$+Log approach.

Time Buckets: We store the history of the graph in a sequence of temporally disjoint time buckets s.t. each time bucket is a logical container of M time windows and their corresponding checkpoints. That is, a checkpoint can be either a delta or a snapshot. Now, we store a snapshot that is valid at the highest time instant of the last time window of a each bucket, whereas we store a delta at the ending time instant of other time windows. Besides, the first $M/2$ time windows are constructed in a forward fashion and ends each with a forward delta. Whereas, the rest of the time windows are constructed in a backward fashion and ends each with a backward delta.

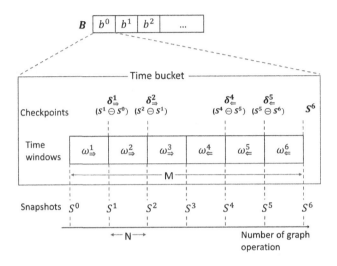

Fig. 13. The internals of the Copy+Log and δ-Copy+Log showing a time bucket (b^0) with $M = 6$, forward and backwards time windows $(\omega^i_{\Rightarrow}, \omega^i_{\Leftarrow})$, deltas $(\delta^i_{\Rightarrow}, \delta^i_{\Leftarrow})$ and snapshot S^i

Time Windows: We use time windows as physical containers for sets of N graph operations each. There are two types of time windows: forward and backward. A forward time window $\omega^i_{\Rightarrow}$ has graph operations sorted in ascending order of their timestamps, while a backward time window $\omega^i_{\Leftarrow}$ has operations sorted in decreasing order of their timestamps after they have been reversed.

Snapshots: A snapshot represents a valid state at the end of the last time window within a bucket. For node or relationship labels, the snapshot includes all existing nodes and relationships at the snapshot time. For dynamic properties, the snapshot includes all nodes and relationships with that property and their latest value before or at the snapshot time.

Deltas: A delta is defined as the minimum number of graph updates required to transform snapshot S into snapshot S'. In other words, if a graph entity is both added and subsequently deleted, these operations will cancel each other out and will not be included in the delta.

Bloom Filter: Bloom filters are assigned to each delta to mitigate the execution time overhead of queries induced by the storage of deltas instead of snapshots. For each graph operation in a delta, we add the identifier of the corresponding node to the Bloom filter. Having this, queries are accelerated by skipping the retrieval of graph operations related to the requested node if the identifier of the latter is not found in the Bloom filter.

6.1   Space and Time Complexity Analysis

This section analyzes the space and time complexities of δ-*Copy+Log*, *Log*, and *Copy+Log* methods, taking into account system and graph parameters such as γ, N, M, c_1, c_2, r_1, r_2, and p_d. These parameters respectively correspond to the set of all graph operations, the number of graph operations in a time window, the number of time windows in a bucket, the size of a single graph operation or element, the time taken to read a graph operation or element, and the probability of deleting a graph element. Note that due to space limitations, this section presents some formulas without detailed explanations of their derivation. A more comprehensive complexity analysis is present in our previous paper [31].

The space usage of the δ-*Copy+Log* is the sum of the space occupied by graph operations, deltas and snapshots. The total space usage of the δ-*Copy+Log* method ($\chi_{\delta-CL}$) can be formulated as follows:

$$\chi_{\delta-CL} = \left(1 + (1 - 2p_d)\frac{(M-2)}{M}\right)c_1|\gamma| + \frac{(1-2p_d)}{2NM}c_2|\gamma|^2$$

The space usage of the *Log* approach (χ_{Log}) is equal to the space occupied by all graph operations (χ_o) which implies the following:

$$\chi_{Log} = c_1|\gamma|$$

The space usage of the *Copy+Log* method (χ_{CL}) is equal to the space occupied by graph operations and snapshots ($\chi_o + \chi_s$) where $M = 1$. Having this, we derive the following:

$$\chi_{CL} = c_1|\gamma| + \frac{(1-2p_d)}{2N}c_2|\gamma|^2$$

From the obtained equations for χ_{Log}, $\chi_{\delta-CL}$ and χ_{CL}, we can derive the following:

$$\chi_{Log} \leq \chi_{\delta-CL} \leq \chi_{CL}$$

We analyze the time complexity of a simplified version of the expand operator for point-based queries. Note that point-based queries are those addressing a single graph snapshot. The expand operator ($\uparrow_\tau (v)$) retrieves all the relationships of a node v whose validity intervals contain the time instant τ.

Execution time of the expand operator: In our analysis, we consider the worst-case execution time of the operator, which involves reading from the snapshot whose timestamp is closest to τ. This, in turn, requires reading all operations in the deltas of the selected time bucket whose time interval is before τ, resulting in the reading of $((\frac{M}{2} - 1)N)$ graph operations where M and N refer to the number of time windows between snapshots and the number of graph operations in each time window, respectively. Finally, we need to read all the graph operations in the time window that follows the last selected delta. Based on this, we derive the following:

$$T_{\delta-CL}\left(\uparrow_\tau(v)\right) = \left(r_2 + \left(\frac{M}{2} - 1\right)Nr_1 + Nr_1\right)$$

The expansion of a node using the Log method might incur loading all graph operations in γ. Having this, we derive the following:

$$T_{Log}\left(\uparrow_\tau(v)\right) = |\gamma|r_1$$

Finally, the expansion of a node using the Copy+Log method incur a single snapshot read which implies the following:

$$T_{Copy+Log}\left(\uparrow_\tau(v)\right) = r_2$$

Consider $|\gamma| \gg (\frac{NM}{2})$ and $|\gamma| \gg \frac{r_2}{r_1}$, then we can derive the following:

$$T_{Copy+Log}\left(\uparrow_\tau(v)\right) \leq T_{\delta-CL}\left(\uparrow_\tau\right)(v) \leq T_{Log}\left(\uparrow_\tau\right)(v)$$

This analysis validates that δ-*Copy+Log* presents a compromise between the *Log* and *Copy+Log* methods. We specifically emphasize analyzing the time complexity of the expand operator as the basis for comparing the time complexity of the δ-*Copy+Log* approach with traditional methods such as *Log* and *Copy+Log*.

7   Overview

This section provides the details of the integration of our temporal graph query language, query processor, and storage technique into Clock-G. Hence, we present in the following the different components composing the architecture of Clock-G (Fig. 14).

Request Handler. The request handler is responsible for managing the read and write requests. As presented in Fig. 14, the request handler comprises two functional components: Reader and Writer.

Our proposed language, T-Cypher, requires the reader to process temporal graph queries using three components: query extractor, planner, and evaluator. The query extractor converts T-Cypher queries into a system-recognized query object using our proposed T-Cypher grammar. The query planner uses our plan

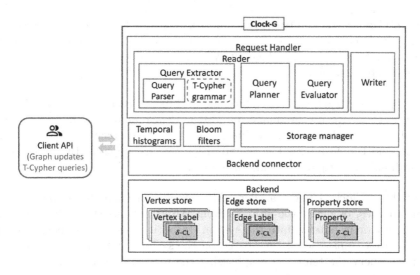

Fig. 14. Overview of the system architecture of Clock-G

selection algorithm to convert the query object into an execution plan that minimizes estimated cardinality of sub-results. Cardinality estimation is based on a cost model and temporal histograms. The query evaluator executes operators using a pool of atomic executors that share intermediate results.

The writer is responsible for inserting graph updates, which involves informing the storage manager which holds the meta data of the δ-*Copy+Log* technique of the insertion. Then, the writer sends an insertion request to the backend connector, which translates the request into atomic write operations for the backend store.

Backend Store. The backend store is responsible for storing the temporal graphs following our proposed storage technique the δ-*Copy+Log*. We rely on the column-oriented database Apache Cassandra [29] for robustness, engineering maturity, and scalability. Besides, Cassandra sorts blocks of data according to a given column or combination of columns. We utilize this feature to sort graph updates according to their chronological order, accelerating their sequential read.

For instance, the storage is separated based on the graph entity type, resulting in node, relationship, and dynamic property stores. For each node/relationship label or dynamic property, we partition the storage based on a Hash partitioning strategy. Each of these partitions corresponds to a storage unit and is stored following the δ-*Copy+Log* method (denoted δ-CL in Fig. 14 for simplicity).

Storage Manager. The storage manager is responsible for applying the rules of the δ-*Copy+Log* method to the storage. Besides, it maintains metadata that helps direct read or write operations to the corresponding storage entities.

Backend Connector. The backend connector connects to and executes requests against the backend store. Hence, it receives read or write requests from the request handler and converts them into Cassandra queries before executing them against the backend store.

Auxiliary Data Structures. To reduce the prohibitive cost of accessing the secondary storage, we use auxiliary data structures maintained in memory and queried when needed. These auxiliary data structures include Temporal histograms and Bloom filters. The temporal histograms represent the evolution of the cardinality of graph elements through time.

Client API. Clock-G offers a client API enabling a client to connect, ingest graph updates, or query the stored graphs. Users can insert graph operations individually or in batches into the system. In both cases, graph operations are attached to a transactional time based on the system's internal clock. Besides, users can query the temporal graph using the temporal graph query language T-Cypher.

8   Evaluation

This section evaluates the performance of Clock-G, aiming to demonstrate that *δ-Copy+Log* provides a balance between the traditional methods *Copy+Log* and *Log* in terms of space usage and evaluation time. It also shows that Clock-G can be tuned to account for acceptable query latency and storage resources. This section also confirms the cost-effectiveness of our query optimizer for T-Cypher queries.

8.1   Experimental Setup

Machine Configuration. The experiments were conducted on a single machine equipped with 32 Intel(R) Xeon(R) E5-2630L v3 1.80GHz CPUs, 264 GB memory, 1 TB SSD, running 64-bit Ubuntu 18.04.4 LTS with 5.0.0-23-generic Linux kernel. We use OpenJDK 11.0.9, Go 1.14.4, DSE 6.8.4, CQL spec 3.4.5 and Neo4j[6] 4.4.

Datasets. We evaluated our proposed methods on synthetic and real temporal graphs to validate their performance. Synthetic datasets were generated with varying probabilities of addition, resulting in three datasets, DS_{p_a}, with p_a values of 0.9, 0.75, and 0.6. These datasets allowed us to analyze the space reduction achieved by *δ-Copy+Log* through the elimination of redundant graph elements across snapshots.

We also used different real-world datasets such as DBLP dataset (DS_{DBLP} [27], Stack overflow dataset (DS_{stack}) and Wiki talk dataset (DS_{wiki}) [44]. To evaluate the sequential paths (presented in Sect. 4), we used the CitiBike dataset[7]

---

[6] https://neo4j.com.

[7] https://ride.citibikenyc.com/system-data.

(DS_{citi}) which includes information of bike trips between stations in New York city.

We present some of the characteristics of the generated datasets in Table 3 where $|V|$ refers to the total number of vertices, $|E|$ refers to the total number of graph operations.

Table 3. Characteristics of the generated graphs

Dataset	$|V|$	$|E|$	Space usage (GB)
DS_{p_a}	500 K	10 M	0.315
DS_{stack}	2.6 M	63.4 M	1.7
DS_{DBLP}	1.8 M	29.5 M	0.831
DS_{wiki}	1.1 M	7.8 M	0.173
DS_{citi}	1 K	2.5 M	0.066
$LDBC_0$	4.2 K	12 K	0.003
$LDBC_1$	406.3 K	1.9 M	0.1
$LDBC_2$	1.1 M	3.9 M	0.3

The storage technique was evaluated using previously described datasets, but a dataset with different relationship and node labels, and time-evolving properties that change over time was required to evaluate complex T-Cypher queries using our query processor. To perform these evaluations, we used the LDBC dataset [9], which represents a temporal social graph where people know each other or like each other's posts and comments. However, the original LDBC schema did not account for dynamic properties, which were necessary for our testing requirements. Thus, we modified the schema by transforming some outgoing relationship types and target nodes into dynamic properties attached to the incident nodes. The modified schema includes nodes for people, posts, and comments, with relationships of type likes or knows. Each node and relationship has a starting and ending time instant that defines the boundaries of the validity time interval of each graph entity, as well as a set of dynamic and static properties that characterize it. For example, the property university of a person node was originally a relationship connected to that node and another university node, which we have converted into a dynamic property. Similarly, we converted the relationships connecting a person to a company and a post or comment to a tag into dynamic properties. We present the characteristics of the generated LDBC graphs ($LDBC_0$, $LDBC_1$, and $LDBC_2$) in Table 3.

8.2 Space Usage and Query Evaluation Time of Basic Temporal Queries

We evaluate disk space usage and query execution time with different system parameter configurations using basic temporal queries such as **local**, **global**,

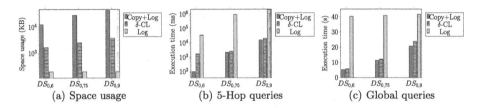

Fig. 15. Comparison with state-of-the-art techniques

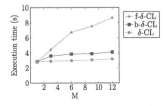

Fig. 16. Evaluation of 8 Hop queries with f-δ-CL, b-δ-CL and δ-CL methods on dataset $DS_{0,6}$ with N set to $10K$

point, and **range** queries. It should be noted that **local** and **global** queries request the local neighborhood of a single query and the state of the entire graph, respectively. **Point** and **range** queries retrieve a point or global state that was valid at a single time instant and during a time range, respectively. These queries can be easily written using the T-Cypher's syntax. In this experiment, local queries start from $1k$ randomly selected vertices, while global queries retrieve snapshots of the graph at uniformly chosen time instants within the time span of the datasets. This evaluation focuses solely on storage technique performance, without using a query processor. More complex T-Cypher queries are evaluated in Sect. 8.3.

Comparison with State-of-the-Art Methods. We compare the results of the proposed method δ-*Copy+Log* with those of the traditional methods *Copy+Log* and *Log*. Now, the implementation of *Copy+Log* in Clock-G is fairly straightforward since it consists of setting parameter M to 1, while implementing *Log* involves creating unbounded time windows.

Figures 15(a), 15(b) and 15(c) display the space usage, the execution time of 5-Hops and global queries on datasets $DS_{0,6}$, $DS_{0,75}$ and $DS_{0,9}$. Note that, we set the system parameters N and M to $10k$ and 12, respectively.

The results clearly demonstrate that the proposed δ-*Copy+Log* method provides a balance between the *Log* and *Copy+Log* approaches. It reduces space usage by a factor of 12 compared to *Copy+Log*, and query execution time by a factor of 340 compared to *Log* for $DS_{0,75}$.

Validating the Use of Bloom Filters. In this evaluation test, we compare the execution time using 3 methods namely: f-δ-CL, b-δ-CL and δ-CL. The f-δ-CL method follows the same approach as the δ-*Copy+Log* with the difference

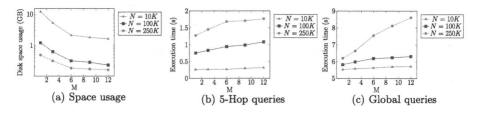

(a) Space usage   (b) 5-Hop queries   (c) Global queries

Fig. 17. Evaluation of the disk space usage and execution time of queries while varying the system's configuration parameter N. The evaluation is conducted on the synthetic dataset $DS_{0,6}$

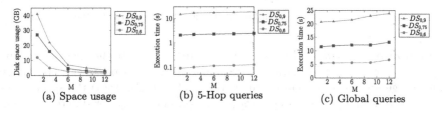

(a) Space usage   (b) 5-Hop queries   (c) Global queries

Fig. 18. Evaluation of the disk space usage and execution time of queries with $N = 10K$. The evaluation is conducted on synthetic datasets $DS_{0,6}$, $DS_{0,75}$ and $DS_{0,9}$ having each a different value of parameter p_a

of storing only forward time windows and deltas and omitting the use of Bloom filters. The b-δ-CL, standing for bloomed-δ-*Copy+Log*, consists of adding Bloom filters to the f-δ-CL. Finally, the δ-CL refers the δ-*Copy+Log* method, hence, consists of adding forward and backward time windows and deltas to the b-δ-CL. Comparing these methods emphasizes the gain of adding Bloom filters and that of storing backward time windows and deltas, separately. Figure 16 shows the execution time of traversal queries with fixed depth 8 on dataset $DS_{0,6}$ while increasing the system parameter M from 1 to 12. The f-δ-CL method significantly increases the execution time with increasing M, but adding Bloom filters to the b-δ-CL reduces the execution time, with a speedup of 52% for $M = 12$. Adding forward and backward time windows and deltas to the δ-CL speeds up traversals by 23% compared to the b-δ-CL. The f-δ-CL method has an overhead of 206% when M is increased from 1 to 12, which is reduced to 12.5% when using the δ-CL.

Variation of N and M. In this evaluation test, we study the effect of system parameters N and M on disk space usage and query execution time. Figure 17(a) shows that increasing M while fixing N significantly reduces space usage compared to the *Copy+Log* method. Smaller values of N result in higher space usage of checkpoints. Increasing M induces more significant disk space gain for smaller values of N. The execution time of 5-Hop and global queries is also evaluated for different configurations of N and M, with results shown in Figs. 17(b) and 17(c).

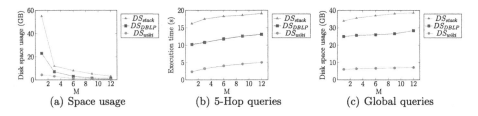

(a) Space usage   (b) 5-Hop queries   (c) Global queries

Fig. 19. Evaluation of the disk space usage and execution time of queries with $N = 250K$. The evaluation is conducted on real datasets DS_{stackO}, DS_{DBLP} and DS_{wiki}

A higher value of N results in higher execution time because fewer checkpoints are created.

Variation of p_a and M. In this evaluation test, we study the effect of varying the linkage probability of datasets DS_{p_a} and the system parameter M on the space usage and query execution time. We display in Fig. 18(a) the space occupied by checkpoints for datasets $DS_{0,6}$, $DS_{0,75}$, and $DS_{0,9}$ for different system configurations where M ranges from 1 to 12. Our results indicate that increasing M leads to a decrease in space usage, and graphs with higher probability of additions provide better space gains. This is because snapshots of such graphs consume more space, making the replacement with deltas more significant in terms of space gain. We also analyze the impact of varying p_a and M on the execution time of 5-Hop and global queries, as shown in Figs. 18(b) and 18(c). We find that increasing p_a leads to an increase in query execution time, as higher node degrees result in more computations to evaluate query results.

Evaluation on Real Datasets. We assess the space efficiency of ingesting real-world datasets using the δ-$Copy$+Log method, with results shown in Fig. 19(a). The space usage of checkpoints created by ingesting datasets DS_{stack}, DS_{DBLP}, and DS_{wiki} into Clock-G reduces significantly when increasing the value of M from 1 to 12. Furthermore, we evaluate 5-Hop traversal and global queries on these real-world datasets, and the results in Figs. 19(b) and 19(c) demonstrate that our solution significantly reduces space usage while adding only a slight query execution time overhead, as compared to the $Copy$+Log method.

Comparison with a Non-temporal Graph Database. In this study, we compare the performance of Clock-G with a commercial graph database Neo4j. To enable the storage and evaluation of temporal graphs in Neo4j, we created a temporal layer by adding validity intervals to each node and relationship occurrence. We tested two implementations: Neo4j without indexes and Neo4j$_i$ with indexes where we add indexes to the starting and ending time instants (***tStart***

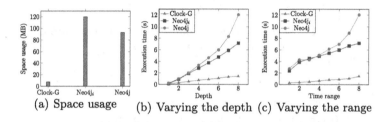

Fig. 20. Evaluation of the space usage and query execution time of Clock-G, Neo4j and Neo4j$_i$

and **tEnd**) of nodes, relationships, and properties[8]. We ingested dataset DS_{citi} in all three systems and evaluated a time increasing path query for each node and for depths 1 to 8 and time ranges 1 hour to 8 hours.

Figure 20(a) compares the space usage of Clock-G with those of Neo4j and Neo4j$_i$. It is evident from the figures that Clock-G consumes less space than Neo4j and Neo4j$_i$.

The Figs. 20(b) and 20(c) show the performance of time increasing path queries with varying depth and time range, comparing the execution time of Clock-G with that of Neo4j and Neo4j$_i$. Results show that Clock-G performs better than the alternative solutions, especially with increasing depth and time range. Clock-G uses parallelism to compute query results and trims the search space to the requested time interval, which is not possible with Neo4j and Neo4j$_i$.

The experiment results emphasize the importance of developing v a graph management system that natively supports temporal data, rather than relying on a non-temporal commercial system.

8.3   Query Execution Time of Complex Temporal Queries

In this section, we evaluate the performance of our query processor by presenting the execution time of T-Cypher queries with the best, random, and worst plan selection. The best execution plan is selected using a greedy algorithm presented in Algorithm 1. For the worst execution plan, a modified version of this algorithm is used where the most expensive algebraic operator is selected at each iteration. Similarly, a random plan is computed using the same algorithm, but the algebraic operator is chosen randomly at each iteration.

Queries. We ran these tests with several T-Cypher queries listed below. Note that all these queries apply to a time interval covering the full history of the LDBC datasets.

---

[8] We use the built-in Neo4j's indexing utility to include indexes on the properties *tStart* and *tEnd*.

```
Query Q0
RANGE_SLICE [2009-01-01T08:00:00Z; 2020-01-01T10:00:00Z]
MATCH (p1:person) -[k1:knows]-> (p2:person)
RETURN p1, p2

Query Q1
RANGE_SLICE [2009-01-01T08:00:00Z; 2020-01-01T10:00:00Z]
MATCH (p1:person) -[k:knows*2]-> (p2:person)
RETURN p1, k, p2

Query Q2
RANGE_SLICE [2009-01-01T08:00:00Z; 2020-01-01T10:00:00Z]
MATCH (p1:person) -[k:knows*3]-> (p2:person)
RETURN p1, k, p2

Query Q3
RANGE_SLICE [2009-01-01T08:00:00Z; 2020-01-01T10:00:00Z]
MATCH (p1:person) -[k1:knows]-> (p2:person)
-[k2:knows]-> (p3:person)
WHERE p1.university=x AND p1@T STARTS k1@T AND p1@T STARTS k2@T
RETURN p1, p2, p3

Query Q4
RANGE_SLICE [2009-01-01T08:00:00Z; 2020-01-01T10:00:00Z]
MATCH (p1:person) -[k1:knows]-> (p2:person) -[k2:knows]->
(p3:person), (p2:person) -[l:likes]-> (p:post)
RETURN p1, p2, p3, p

Query Q5
RANGE_SLICE [2009-01-01T08:00:00Z; 2020-01-01T10:00:00Z]
MATCH (p1:person) -[k:knows]-> (p2:person) -[l1:likes]->
(p:post), (p1:person) -[l2:likes]-> (p:post)
RETURN p1, p2, p
```

Q_0 returns the pairs of persons who knew each other in the time interval. Q_1 returns the person's 2-hop friendship paths. Q_2 returns the person's 3-hop friendship paths. Q_3 returns the friends of friends of a person who went to the university x such that the friendship started when the person studied in that university. Q_4 returns all friends of friends of each person such that the intermediate person likes a post. Q_5 returns the friends who like the same post. Some of these queries include graph entities with varying levels of granularity. For example, the *knows* relationships are more selective than the *likes* relationships. In such cases, it is reasonable for the query processor to prioritize loading the "knows" relationships prior to the "likes" relationships during evaluation in queries $Q4$ and $Q5$.

Plan Selection. This section shows the best, random, and worst execution plans of Query Q_3 due to space constraints. Notably, the dataset exhibits a lower cardinality for the *person* label compared to the *post* label, and a higher cardinality for the *like* relationship than the *know* type.

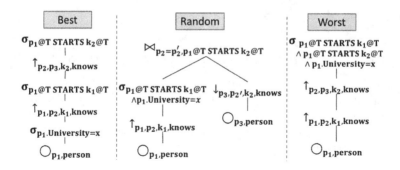

Fig. 21. Best, random, and worst evaluation plan of Query Q_3

We present the evaluation plans for Query Q_3 in Fig. 21. The optimal plan begins by retrieving nodes p_1 and selecting those who attended the specified *university*. It then expands nodes p_1 with relationship k_1, selects those who began at the *university*, expands again with relationship k_2, and selects those who began with p_1. As predicted, the worst plan postpones selections until the end. The depicted random plan computes two subparts of the query and joins the results. The first subpart retrieves nodes p_1 and their direct neighbors p_2 connected via relationship k_1, selecting only those who studied at the given *university*. The second subpart retrieves nodes p_3 and their direct neighbors p_2' connected via relationship k_2. The two subparts are then joined based on the condition that p_2 should equal p_2' and the temporal condition that k_2 should be started by p_1.

Figure 22 displays the average execution time resulting from the best, random, and worst plan selection strategies for computing queries $\{Q_0, \ldots, Q_5\}$ on datasets $LDBC_0$, $LDBC_1$, and $LDBC_2$. Note that these results correspond to the average computation time of 10 repetitions.

The execution time of queries Q_0, Q_1, and Q_2 increases with the number of traversed hops for all evaluated datasets, but the best, random, and worst evaluation plans show negligible differences. This difference is expected, as all node and relationship variables in these queries share the same label and type, resulting in the same cardinality. In contrast, the execution time for queries Q_3, Q_4, and Q_5 differs significantly between the best and worst evaluation plans since these queries present graph entities of different granularity.

The worst plan selection strategy delays the selection process until the end and begins with the nodes having the highest cardinalities. This negatively impacts the cost of query evaluation. On the other hand, selecting plans randomly provides a balance between the best and worst plan selection strategies in terms of query execution time. This is because there are several alternative query plans that fall between the best and worst plan selections. Therefore,

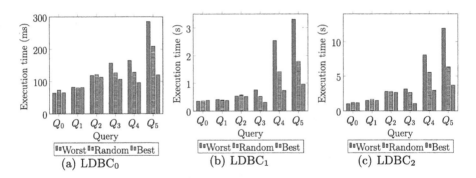

Fig. 22. Comparison between the execution time of T-Cypher queries with worst, random, and best execution plans

these results demonstrate the effectiveness of our plan selection algorithm and cost model.

Comparison with Neo4j. We compared Clock-G with a non-temporal graph system by introducing a temporal layer on top of Neo4j. This layer handles the temporal dimension by storing time instants for graph updates and converting T-Cypher queries into Cypher queries.

We compared the performance of Neo4j and Clock-G in executing queries $\{Q_0, \ldots, Q_5\}$ using Algorithm 1 and present the results in Fig. 23. Clock-G outperforms Neo4j by up to 80% due to its ability to prune the search space and directly search within selected time windows, snapshots, and deltas.

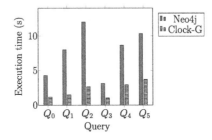

Fig. 23. Comparison between the execution time of T-Cypher queries with Neo4j and Clock-G on LDBC$_2$

9   Conclusion

In this paper, we presented Clock-G, a temporal graph management system with a holistic approach to covering query language, processing, and storage. T-Cypher is our user-friendly query language that allows for temporal constraints on graph pattern matching and navigational queries. Our query processor evaluates T-Cypher queries and targets the minimization of the processing cost. To address this, our processor uses a graph algebra that defines the algebraic operators of a plan, cost model that defines the cost of each operator, and temporal histograms that preserve the cardinality's evolution. Our storage technique, δ-*Copy+Log*, targets the reduction of the space usage of the traditional *Copy+Log* technique by storing deltas instead of full graph snapshots. Tests on synthetic and real-world graphs show that δ-*Copy+Log* significantly reduces space usage and execution time compared to traditional methods and validates the efficiency of our query processor.

A promising direction for future work is the incorporation of the spatial dimension into Clock-G, as it has been the subject of research in the area of spatio-temporal databases [15,36]. To further enhance the capabilities of the Thing'in platform, we can explore spatio-temporal queries, such as expressing a geographic region in which objects should be located during a specified time interval.

References

1. Allen, J.F.: Maintaining knowledge about temporal intervals. Commun. ACM **26**(11), 832–843 (1983)
2. Arenas, M., Bahamondes, P., Aghasadeghi, A., Stoyanovich, J.: Temporal regular path queries. In: 2022 IEEE 38th International Conference on Data Engineering (ICDE), pp. 2412–2425 (2022). https://doi.org/10.1109/ICDE53745.2022.00226
3. de Berg, M., van Kreveld, M., Overmars, M., Schwarzkopf, O.: Computational geometry. In: Computational Geometry, pp. 1–17. Springer, Heidelberg (1997). https://doi.org/10.1007/978-3-662-03427-9_1
4. Castelltort, A., Laurent, A.: Representing history in graph-oriented NOSQL databases: a versioning system. In: Eighth International Conference on Digital Information Management (ICDIM 2013), pp. 228–234. IEEE (2013)
5. Cattuto, C., Quaggiotto, M., Panisson, A., Averbuch, A.: Time-varying social networks in a graph database: a neo4j use case. In: First International Workshop on Graph Data Management Experiences and Systems. GRADES 2013, New York, NY, USA. Association for Computing Machinery (2013). https://doi.org/10.1145/2484425.2484442, https://doi.org/10.1145/2484425.2484442
6. Clifford, C.S.J.J., et al.: A consensus glossary of temporal database concepts. SIGMOD Record **23**(1), 52–64 (1994)
7. Debrouvier, A., Parodi, E., Perazzo, M., Soliani, V., Vaisman, A.: A model and query language for temporal graph databases. VLDB J. **30**(5), 825–858 (2021). https://doi.org/10.1007/s00778-021-00675-4
8. Deutsch, A., et al.: Graph pattern matching in GQL and SQL/PGQ. In: Proceedings of the 2022 International Conference on Management of Data. SIGMOD 2022, New York, NY, USA, pp. 2246–2258. Association for Computing Machinery (2022). https://doi.org/10.1145/3514221.3526057
9. Erling, O., et al: The LDBC social network benchmark: interactive workload. In: Proceedings of the 2015 ACM SIGMOD International Conference on Management of Data. SIGMOD 2015, New York, NY, USA, pp. 619–630. Association for Computing Machinery (2015). https://doi.org/10.1145/2723372.2742786

10. Francis, N., et al.: Cypher: an evolving query language for property graphs. In: Proceedings of the 2018 International Conference on Management of Data, pp. 1433–1445 (2018)
11. George, B., Kang, J.M., Shekhar, S.: Spatio-temporal sensor graphs (STSG): a data model for the discovery of SPATIO-temporal patterns. Intell. Data Anal. **13**(3), 457–475 (2009)
12. Ghrab, A., Romero, O., Skhiri, S., Vaisman, A.A., Zimányi, E.: GRAD: on graph database modeling. CoRR abs/1602.00503 (2016). http://arxiv.org/abs/1602.00503
13. Grandi, F.: T-SPARQL: A TSQL2-like temporal query language for RDF. In: ADBIS (Local Proceedings). pp. 21–30. Citeseer (2010)
14. Gubichev, A.: Query processing and optimization in graph databases. Ph.D. thesis, Technische Universität München (2015)
15. Hadjieleftheriou, M., Kollios, G., Gunopulos, D., Tsotras, V.J.: On-line discovery of dense areas in spatio-temporal databases. In: Hadzilacos, T., Manolopoulos, Y., Roddick, J., Theodoridis, Y. (eds.) SSTD 2003. LNCS, vol. 2750, pp. 306–324. Springer, Heidelberg (2003). https://doi.org/10.1007/978-3-540-45072-6_18
16. Haeusler, M., Trojer, T., Kessler, J., Farwick, M., Nowakowski, E., Breu, R.: ChronoGraph: a versioned TinkerPop graph database. In: Filipe, J., Bernardino, J., Quix, C. (eds.) DATA 2017. CCIS, vol. 814, pp. 237–260. Springer, Cham (2018). https://doi.org/10.1007/978-3-319-94809-6_12
17. Han, W., Li, K., Chen, S., Chen, W.: Auxo: a temporal graph management system. Big Data Min. Anal. **2**(1), 58–71 (2018)
18. Han, W., et al.: Chronos: a graph engine for temporal graph analysis. In: Proceedings of the Ninth European Conference on Computer Systems, pp. 1–14 (2014)
19. Hartig, O., Pérez, J.: Semantics and complexity of GRAPHQL. In: Proceedings of the 2018 World Wide Web Conference. WWW 2018, International World Wide Web Conferences Steering Committee, Republic and Canton of Geneva, CHE, pp. 1155–1164 (2018). https://doi.org/10.1145/3178876.3186014
20. Hartmann, T., Fouquet, F., Jimenez, M., Rouvoy, R., Le Traon, Y.: Analyzing complex data in motion at scale with temporal graphs (2020)
21. Huo, W., Tsotras, V.J.: Efficient temporal shortest path queries on evolving social graphs. In: Proceedings of the 26th International Conference on Scientific and Statistical Database Management. SSDBM 2014, New York, NY, USA. Association for Computing Machinery (2014). https://doi.org/10.1145/2618243.2618282
22. Hölsch, J., Grossniklaus, M.: An algebra and equivalences to transform graph patterns in neo4j. In: Palpanas, T., Stefanidis, K. (eds.) Proceedings of the Workshops of the EDBT/ICDT 2016 Joint Conference (EDBT/ICDT 2016). No. 1558 in CEUR Workshop Proceedings (2016). http://ceur-ws.org/Vol-1558/paper24.pdf
23. Kempe, D., Kleinberg, J., Kumar, A.: Connectivity and inference problems for temporal networks. J. Comput. Syst. Sci. **64**(4), 820–842 (2002). https://doi.org/10.1006/jcss.2002.1829, https://www.sciencedirect.com/science/article/pii/S0022000002918295
24. Khurana, U., Deshpande, A.: Efficient snapshot retrieval over historical graph data. In: 2013 IEEE 29th International Conference on Data Engineering (ICDE), pp. 997–1008. IEEE (2013)
25. Khurana, U., Deshpande, A.: Storing and analyzing historical graph data at scale. arXiv preprint arXiv:1509.08960 (2015)
26. Koloniari, G., Souravlias, D., Pitoura, E.: On graph deltas for historical queries. arXiv preprint arXiv:1302.5549 (2013)

27. Kunegis, J.: Konect: The koblenz network collection. WWW 2013 Companion, New York, NY, USA, pp. 1343–1350. Association for Computing Machinery (2013). https://doi.org/10.1145/2487788.2488173

28. Labouseur, A.G., et al.: The g* graph database: efficiently managing large distributed dynamic graphs. Distrib. Parallel Databases **33**(4), 479–514 (2015)

29. Lakshman, A., Malik, P.: Cassandra: a decentralized structured storage system. ACM SIGOPS Oper. Syst. Rev. **44**(2), 35–40 (2010)

30. Macko, P., Marathe, V.J., Margo, D.W., Seltzer, M.I.: Llama: efficient graph analytics using large multiversioned arrays. In: 2015 IEEE 31st International Conference on Data Engineering, pp. 363–374. IEEE (2015)

31. Massri, M., Miklos, Z., Raipin, P., Meye, P.: Clock-g: a temporal graph management system with space-efficient storage technique. In: 2022 IEEE 38th International Conference on Data Engineering (ICDE), Los Alamitos, CA, USA, pp. 2263–2276. IEEE Computer Society(2022). https://doi.org/10.1109/ICDE53745.2022.00215, https://doi.ieeecomputersociety.org/10.1109/ICDE53745.2022.00215

32. Miao, Y., et al.: Immortalgraph: a system for storage and analysis of temporal graphs. ACM Trans. Storage (TOS) **11**(3), 1–34 (2015)

33. Moffitt, V.Z., Stoyanovich, J.: Temporal graph algebra. DBPL 2017, New York, NY, USA. Association for Computing Machinery (2017). https://doi.org/10.1145/3122831.3122838

34. Montanari, A., Chomicki, J.: Time domain. In: Liu, L., Özsu, M.T. (eds.) Encyclopedia of Database Systemspp, pp. 3103–3107. Springer, Boston (2009). https://doi.org/10.1007/978-0-387-39940-9_427

35. Pan, R.K., Saramäki, J.: Path lengths, correlations, and centrality in temporal networks. Phys. Rev. E **84**(1), 016105 (2011)

36. Perry, M., Jain, P., Sheth, A.P.: SPARQL-ST: extending SPARQL to support spatiotemporal queries. In: Ashish, N., Sheth, A. (eds.) Geospatial Semantics and the Semantic Web. Semantic Web and Beyond, vol. 12, pp. 61–86. Springer, Boston (2011). https://doi.org/10.1007/978-1-4419-9446-2_3

37. Ramesh, S., Baranawal, A., Simmhan, Y.: A distributed path query engine for temporal property graphs. In: 2020 20th IEEE/ACM International Symposium on Cluster, Cloud and Internet Computing (CCGRID), pp. 499–508 (2020). https://doi.org/10.1109/CCGrid49817.2020.00-43

38. Redmond, U., Cunningham, P.: Subgraph isomorphism in temporal networks. CoRR abs/1605.02174 (2016). http://arxiv.org/abs/1605.02174

39. Rizzolo, F., Vaisman, A.A.: Temporal xml: modeling, indexing, and query processing. VLDB J.-Int. J. Very Large Data Bases **17**(5), 1179–1212 (2008)

40. Rost, C., et al.: Distributed temporal graph analytics with gradoop. VLDB J. **31**(2), 375–401 (2022). https://doi.org/10.1007/s00778-021-00667-4

41. Semertzidis, K., Pitoura, E., Lillis, K.: Timereach: historical reachability queries on evolving graphs. In: EDBT (2015)

42. Semertzidis, K., Pitoura, E., Terzi, E., Tsaparas, P.: Best friends forever (BFF): finding lasting dense subgraphs. CoRR abs/1612.05440 (2016). http://arxiv.org/abs/1612.05440

43. Snodgrass, R.T., et al.: TSQL2 language specification. SIGMOD Rec. **23**(1), 65–86 (1994). https://doi.org/10.1145/181550.181562

44. Wen, D., Huang, Y., Zhang, Y., Qin, L., Zhang, W., Lin, X.: Efficiently answering span-reachability queries in large temporal graphs. In: 2020 IEEE 36th International Conference on Data Engineering (ICDE), pp. 1153–1164 (2020). https://doi.org/10.1109/ICDE48307.2020.00104

45. Wu, H., Huang, Y., Cheng, J., Li, J., Ke, Y.: Reachability and time-based path queries in temporal graphs. In: 2016 IEEE 32nd International Conference on Data Engineering (ICDE), pp. 145–156 (2016). https://doi.org/10.1109/ICDE.2016.7498236

46. Xiangyu, L., Yingxiao, L., Xiaolin, G., Zhenhua, Y.: An efficient snapshot strategy for dynamic graph storage systems to support historical queries. IEEE Access **8**, 90838–90846 (2020)

47. Zhang, T., Gao, Y., Qiu, L., Chen, L., Linghu, Q., Pu, S.: Distributed time-respecting flow graph pattern matching on temporal graphs. World Wide Web **23**(1), 609–630 (2020). https://doi.org/10.1007/s11280-019-00674-0

TSPredIT: Integrated Tuning of Data Preprocessing and Time Series Prediction Models

Rebecca Salles[1], Esther Pacitti[2,3], Eduardo Bezerra[1], Celso Marques[1],
Carla Pacheco[1], Carla Oliveira[1,3,4], Fabio Porto[4], and Eduardo Ogasawara[1(✉)]

[1] Federal Center for Technological Education of Rio de Janeiro (CEFET/RJ), Rio de
Janeiro, Brazil
{rebecca.salles,carla.pacheco}@eic.cefet-rj.br,
{ebezerra,celso.silva}@cefet-rj.br, carla.oliveira@ibge.gov.br,
eogasawara@ieee.org
[2] University of Montpellier, Montpellier, France
Esther.Pacitti@lirmm.fr
[3] National Institute for Research in Digital Science and Technology (INRIA),
University of Montpellier, Montpellier, France
[4] National Laboratory for Scientific Computing (LNCC), Petropolis, Brazil
fporto@lncc.br

Abstract. Prediction is one of the most important activities while working with time series. There are many alternative ways to model the time series. Finding the right one is challenging to model them. Most data-centric models (either statistical or machine learning) have hyperparameters to tune. Setting them right is mandatory for good predictions. It is even more complex since time series prediction also demands choosing a data preprocessing that complies with the chosen model. Many time series frameworks, such as Scikit Learning, have features to build models and tune their hyperparameters. However, only some works address tuning data preprocessing hyperparameters and model building. TSPredIT addresses this issue in this scope by providing a framework that seamlessly integrates data preprocessing activities with models' hyperparameters. TSPredIT is made available as an R-package, which provides functions for defining and conducting time series prediction, including data pre(post)processing, decomposition, hyperparameter optimization, modeling, prediction, and accuracy assessment. Besides, TSPredIT is also extensible, which significantly expands the framework's applicability, especially with other languages such as Python.

Keywords: time series · prediction · data preprocessing · machine learning · hyperparameter optimization

1 Introduction

The prediction of time series has gained more attention in the last decades. Many time series prediction methods have been developed and can be found

© Springer-Verlag GmbH Germany, part of Springer Nature 2023
A. Hameurlain et al. (Eds.): *TLDKS LIV*, LNCS 14160, pp. 41–55, 2023.
https://doi.org/10.1007/978-3-662-68014-8_2

in the literature [4]. An adequate prediction method is mandatory for building the right model [31], especially for data-driven models. They are generally organized between statistical and machine learning [20]. These types of methods usually have to set hyperparameters. In this sense, hyperparameter optimization is a fundamental step since it can influence the predictive performance of the resulting models [16,19].

Additionally, these models might be improved by adequate data preprocessing activities. Most of these methods tend to be optimistic regarding their assumptions over the time series and are not ready to handle nonstationarity [4,30]. They also suffer from the presence of concept drift [21] or lack of data [33]. These two cases are related. When concept drift occurs, generally, there are few samples to support model building. Usually, nonstationarity demands transformation methods to address this issue [30]. Besides, while working with small samples, data augmentation techniques are also needed [23,33,35].

For the algorithm to make predictions with greater accuracy, optimizing the hyperparameters is necessary [34]. Hyperparameters are values that make up the initial configuration of the learning algorithm [9]. Hyperparameters are also present in data preprocessing methods. Several factors influence the predictive performance of time series models, mainly choosing and tuning the right methods for data preprocessing and hyperparameters.

Regarding statistical learning, some methods seamlessly integrate hyperparameters optimization of data processing techniques. It includes the autoregressive integrated moving average (`ARIMA`) algorithm that optimize parameters (p, d, q) for `ARIMA` [12]. The d parameter represents the Integrated part of ARIMA and performs the differentiation of observations internally as a preprocessing step for the series to be stationary. The p parameter corresponds to the AR part of ARIMA, the number of autoregressive terms. The MA model works with the size of the moving average window and is represented by the q parameter. This method tunes altogether differentiation, autoregressive, and moving average models [3]. Conversely, there are many frameworks for machine learning, such as Scikit learn [11,26], which provides (i) a broad range of prediction methods, (ii) an extensive set of preprocessing methods, (iii) hyperparameter optimization features for machine learning. However, directly optimizing data processing and machine learning is left for users to program according to their needs.

In this context, this paper presents TSPredIT, an evolved version of TSPred [31] that seamlessly integrates the tuning of data preprocessing and time series prediction models for univariate time series. It only concerns regression models and is specialized integrating data transformation methods and data augmentation to aid in building machine learning methods (`MLM`) prediction models. TSPredIT is made available as an R-package. It is the first tool to seamlessly integrate a broad range of data transformation and preprocessing methods and state-of-the-art statistical and machine learning prediction methods for addressing nonstationary time series. The package automates the time series prediction process and parameterization while enabling user-defined prediction methods and data transformations, including code built in other languages like Python. Due to that, the features provided by TSPredIT are shown to be competitive regarding time series prediction accuracy.

Besides this introduction, the paper is organized into five more sections. Section 2 presents the background, while Sect. 3 presents the related work. Section 4 presents the TSPredIT, which evolves from the previous version of TSPred [31]. Section 5 provides a clear example of using TSPredIT's main features. Conversely, Sect. 6 characterizes the effect of choosing data preprocessing and different MLM during prediction. Finally, Sect. 7 concludes the work.

2   Background

Time series prediction is commonly associated with the scenario of regression. For simplicity, the paper may refer to prediction and regression interchangeably. Relevant models adopted for time series prediction generally fall into the categories of statistical or machine learning models [29]. The accuracy of the predictions depends on the quality of the historical data, the appropriateness of the model, and the assumptions made about the underlying processes driving the time series [10,14].

Figure 1 presents a general time series prediction process. It encompasses five main activities. It provides a general framework for predicting a time series based on a particular setup of preprocessing methods and prediction models. They are briefly described here, and some parts are detailed in the following sections.

Fig. 1. Time series prediction process [29]

Activity 1, depicted in Fig. 1 in purple, refers to acquiring the time series and performing data preprocessing. It is generally associated with data cleaning, normalization, and transformation but might include other techniques, such as data augmentation. The transformations commonly change the time series domain values, and their parameters must be stored to support later detransformation to the original domain. For time series prediction, splitting the time series into a training and test set is also important during data preprocessing. All data preprocessing parameters should be computed during training and reapplied from the tune-values of training during the test. The model is built using the training slice and evaluated using the unseen test set, always ahead. However, when the goal is to adjust a model for the time series, the model does not need to be partitioned into a training and test set.

Activity 2, in blue, addresses model training. The prediction methods very often require hyperparameter optimization. In such a case, the training slice is again split into a novel training and validation set. Alternative models exploring hyperparameters values are built using the novel training and evaluated using the validation set. Once hyperparameters are fixed, a single model is built using the entire training dataset. From this moment, the model is available for use.

Activity 3, also in blue, refers to the model prediction. It is worth mentioning that the predicted values are not in the time series domain. In this sense, they can not be directly evaluated. Data postprocessing is needed in this case. Activity 4, also in purple, corresponds to the postprocessing of predictions, reversing transformations applied to the time series data in Activity 1. In a macro view, the data is normalized (scaled) and given as input to an algorithm. After the forecast, a denormalization process maps back the predicted values into the original scale of the time series. An example of postprocessing can be seen in previous work [24], where the data is denormalized to measure the error in the same scale for comparison purposes.

Finally, Activity 5, in pink, is the evaluation of prediction errors yielded by the model, as well as model fitness metrics. If the results are inadequate, this process can be revised and repeated to refine models. This process iteratively improves the quality of predictions (for time series prediction) or model adjustment (for time series modeling). The prediction can be evaluated in several ways, mostly measuring the errors between prediction and actual observation, such as Mean Square Error (MSE) and symmetric MAPE (sMAPE) in a test set. Alternatively, they can be measured by the level of model adjustment, such as Akaike Information Criterion (AIC) and Bayesian Information Criterion (BIC) [31].

This process provides a systematic way of predicting a time series based on particular preprocessing and prediction methods. It also focuses on prediction and model evaluation, that is, evaluating the accuracy of prediction and the fitness of a model. Such evaluation may indicate a demand for refining and perfecting the preprocessing-modeling setup and its parameters to obtain a more accurate model. This process may be repeated if the evaluated time series prediction model does not reach the desired accuracy. This process enables benchmarking different preprocessing-modeling setups.

3   Related Work

Several authors focused on the task of exploring different models. Ramey [27] and Lessmann et al. [18] developed frameworks for evaluating classification models and algorithms. Moreover, Bischl et al. [2] and Eugster and Leisch [8] developed the R-packages *mlr* and *benchmark*, respectively, which provide tools for executing automated experiments when benchmarking a set of models for data mining tasks such as classification and regression. These packages are designed to support tabular data and focus on benchmarking based on plot visualization.

Hyndman and Khandakar [12] and Hyndman et al. [13] present frameworks for automatic forecasting using mainly statistical models such as ARIMA and exponential smoothing state space model (ETS). Hyndman and Khandakar [12] produced the well-known R-package named *forecast*, which can be used for automatic time series prediction. The R-package of Moreno, Rivas, and Godoy [22] also facilitates time series prediction using simple differencing (diff) and Box-Cox transform (BCT). Furthermore, we observed three works worth mentioning. Diebold and Mariano [7] propose various tests to compare the predictive accuracy of two different prediction models. Diebold and Lopez [6] propose an ensemble approach using different prediction models. Kumar et al. [17] propose a class of analytics systems to manage model selection using key ideas from data management research.

Besides, hyperparameters optimization is also a deeply studied subject [1, 15, 16]. The studies may focus on exploring the hyperparameter search space using a certain heuristic. Conversely, some approaches target the right establishing of the hyperparameters to explore using either grid search or a more advanced search strategy. The Grid Search approach is commonly adopted to explore a broad range of hyperparameter settings. It consists of repeatedly training the learning algorithm with different possible hyperparameter settings combinations. At the end of the process, the hyperparameter setting that resulted in the lowest prediction errors (measured in a separate validation set) is chosen [34]. Such optimized hyperparameter settings can then be used to fit the learning model [28]. All these approaches try to lower global prediction error in machine learning but are resilient to the problem of data overfitting. Such an issue occurs when the fitted model is too dependent on the training dataset. One consequence is the fitted model's inability to generalize to unseen data observations [32].

All in all, several works present frameworks and tools for MLM performance assessment. Nonetheless, to our knowledge, no work proposes and implements a framework for the seamless integration of hyperparameter optimization of data preprocessing and time series prediction methods.

4   TSPredIT

The main modules of the TSPredIT framework are depicted in Fig. 2 as a UML class diagram. TSPredIT has five main functionality modules: *Preprocessing* (in purple), *Modeling* (in blue), *Sampling* (in yellow), *Evaluating* (in pink), and *Tuning* (in green). Together, they are used to support the time series prediction process. The colors of the classes are associated with their participation in the time series prediction process as depicted in Fig. 1.

All classes are inherited from TSBase. It provides a basic fit method and some attributes for introspection (to support provenance). It also includes a fit analysis of the data to adjust basic parameter values. A TSData class also provides a uniform perspective for time series data and its transformation to sliding windows. These two types are a specialization of TSData.

The first module is responsible for preprocessing (transform) and postprocessing (inverse transform) a time series. The model groups two main features. The first is related to data transformation. Especially it includes the implementation of the main nonstationary time series transformation methods [30], being either mapping-based, namely the logarithmic transform (LT), BCT, percentage change transform (PCT), moving average smoother (MAS), and diff, or splitting-based, such as empirical mode decomposition (EMD) and wavelet transform (WT). All these methods are implemented as a specialization of the Transformation class. The basic implements fit, transform, and inverse transform to provide the desired behavior.

Furthermore, it also groups a set of methods related to data augmentation. All methods include warping, flipping, and jittering [23,33]. These methods are used during tuning, which is explained later.

The second module is related to Sampling. It is responsible for converting a time series to sliding windows. It is also responsible for separating the data (TSData) into training and testing. The test size uses only recent observations to avoid introducing new data during training. Like augmentation, sampling is used during tuning.

Other relevant preprocessing methods for time series prediction are specialized from Preprocessing class, in purple. It includes support for handling missing values and data normalization such as Sliding Windows min-max normalization (swminmax), Min-max normalization (gminmax), diff, and Adaptive normalization (an) methods [24].

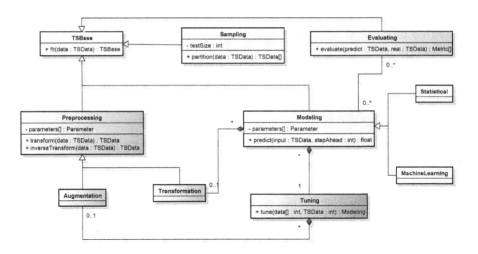

Fig. 2. TSPred-IT functionality modules and pre-implemented algorithms (Color figure online)

The *Modeling* module, in blue, is responsible for modeling (fit) and predicting (predict) a time series based on a particular time series prediction method. These tasks are specialized for either statistical or machine learning models. For the latter, the framework is prepared to perform any necessary machine learning life-cycle tasks during the training and prediction steps, including coercing data into sliding windows, normalizing and transforming the input data. This module includes the implementation of the statistical models: ARIMA, Holt-Winter's exponential smoothing (HW), theta forecasting (TF), and ETS. *MLM* models include multilayer perceptron network (MLP), random forest regression (RFR), support vector machine (SVM), MLP, and extreme learning machines network (ELM). Furthermore, the module provides deep learning models available in the *PyTorch* library, namely convolutional neural network (Conv1D) and long short-term memory neural network (LSTM). Models are associated with zero or one data preprocessing transformation and data augmentation technique. However, it might be set up with multiple candidate options chosen during tuning.

The *Tuning* module is designed to provide hyperparameter optimization. It is invoked during the fitting of a model. Hyperparameter optimization occurs whenever the modeling of time series or preprocessing transformations has a degree of freedom to adjust. The default Tuning applies time series cross-validation using the training set [14]. Data augmentation might be applied in each partition, and the model is trained after applying the data transformation. The fittest model and data preprocessing method were discovered using a grid-search, *i.e.*, the one that leads to better prediction during cross-validation, is chosen for training using the entire training set. Since this is not the only way of conducting Hyperparameter optimization, the default Tuning class can be specialized to provide other ways of enhancing this feature.

Finally, the *Evaluating* module, in pink, is responsible for assessing the model fitness and quality of predictions. These tasks are specialized for computing either prediction accuracy (error) measures or model fitting criteria. The available prediction accuracy measures include MSE, sMAPE, and maximal error. It also includes model fitness criteria such as AIC, BIC, and log-likelihood [5].

TSPredIT can integrate the described modules in a *workflow*, connecting the five modules described. The package provides the means to perform the *benchmarking* of several prediction models. It is important to remark that although providing several pre-implemented options, TSPredIT design enables the user to define and apply customized time series prediction methods.

Moreover, the package provides several automatized features for any time series prediction application. Among them, some of the main features are (i) seamless recursive combination of two or more transformation methods; (ii) seamless integration of transformation methods to the prediction process [30], which demands the combination of predictions for each component resulting from data decomposition (first package to include this approach); (iii) transformation and model parameter selection; (iv) multistep-ahead or one-step-ahead predictions; (v) rolling origin evaluation [14] for both statistical and machine learning models, and (vi) machine-learning life-cycle tasks performed during training and prediction steps. Data normalization and sliding window transformation are seamlessly conducted during machine learning model training.

The framework is implemented in R using the S3 class system [36]. TSPredIT is currently available on GitHub[1]. It is an ongoing evolution of TSPred [31], built on top of the DAL Toolbox[2].

---

[1] https://github.com/cefet-rj-dal/tspredit.
[2] https://cran.r-project.org/web/packages/daltoolbox/index.html.

5   Usage of TSPredIT

This section gives examples of *TSPredIT* usage. The first example corresponds to a time series prediction using a wrapper for the MLP model using sliding windows min-max normalization. The hyperparameter tuning applies a grid search using time series cross-validation.

The Listing 1.1 the *TSPredIT* R-package processes a time series. The lines (1–3) of code target and load the installation of *TSPredIT*. The components for the time series process can be defined separately to enable reuse. Besides, the dataset used is made available in the R-package. It is loaded using the data function (line 6).

The time series is converted into sliding windows (line 8). All sliding windows are shifted with overlap with step 1 by default. In the example, the size of the sliding windows is 8. Besides, the last 4 windows are reserved for testing (line 10), and the complement is used for testing in order to control more precisely which observations is being considered, as a fine tuning. Finally, the training data is separated into input and output (line 12).

The hyperparameter setup is established in lines 15–17. It indicates the data preprocessing option of min-max sliding windows. The input size for model building varies between 3 and 7. The base model is related to MLP. No data augmentation method is used in this example: *ts_augment*(). Also, some specific parameters for MLP are indicated in lines 18–19. It provides ranges for the number of neurons in the hidden layer, the rate of decay during training, and the maximum number of iterations (fixed). In lines 20–21, the actual tuning is executed using the training set. Internally, it splits the data using time series cross-validation. The build model hyperparameters that work better during cross-validation are used to build the final model using the entire training set. In this example, 500 configurations were explored, each one ten times due to the default ten-fold cross-validation.

Lines 23–26 present the level of adjustment for the time series in the training set. This aspect is important since the error level in training is commonly higher during testing. It provides an expected entry error. Lines 28–31 present the prediction for testing. It applies a rolling origin with one step-ahead prediction, leading to four predictions from previously known observations. The sMAPE is presented in line 33.

To clarify how extensible is *TSPredIT* in providing alternative MLM, Listing 1.2 changes four lines of code to switch the MLP to ELM, with different ranges of hyperparameters to explore. This feature is possible due to the wrapper classes provided by TSPredIT that integrate state-of-the-art methods. Additionally, novel methods can be wrapped. Writing the fit and predict methods is needed to incorporate a novel method at *TSPredIT*.

6   Features Evaluation

TSPredIT was experimentally evaluated to expose the main features of the framework. For that, it was derived a time series dataset from public data avail-

Listing 1.1. Example of time series prediction process in TSPredIT

```
1  > library(daltoolbox)
2  #https://cefet-rj-dal.github.io/tspredit/
3  > library(tspredit)
4
5  # Loading fertilizers dataset
6   > data(fertilizers)
7  # Converting to sliding windows
8   > ts <- ts_data(fertilizers$brazil_n, sw = 8)
9  # Partitioning into training and testing
10  > samp <- ts_sample(ts, test_size = 4)
11 # Separeting input and output for training
12  > io_train <- ts_projection(samp$train)
13
14 # Hyperpararameter
15  > tune <- ts_maintune(preprocess = list(
16     ts_norm_swminmax()), input_size = c(3:7),
17     base_model = ts_mlp(), augment = list(ts_aug_none()))
18  > ranges <- list(size = 1:10, decay = seq(0, 1, 1 / 9),
19            maxit = 10000)
20  > model <- fit(tune, x = io_train$input,
21            y = io_train$output, ranges)
22
23 # Measuring the level of adjustment
24  >   adjust <- predict(model, io_train$input)
25  >   ev_adjust <- evaluation.tsreg(io_train$output, adjust)
26  >   print(ev_adjust$metrics$sMAPE)
27
28 # Obtaining the prediction
29  > io_test <- ts_projection(samp$test)
30  > prediction <- predict(model,  x = io_test$input,
31            steps_ahead = 1)
32  > ev_test <- evaluation.tsreg(io_test$output, prediction)
33  > print(ev_test$metrics$sMAPE)
```

Listing 1.2. R example present the simplicity to explore different setups

```
1  # Hyperpararameter
2  > tune <- ts_maintune(preprocess = list(ts_norm_swminmax()),
3          input_size = c(3:7), base_model = ts_elm(),
4          augment = list(ts_aug_none()))
5  > ranges <- list(nhid = 1:20,
6          actfun=c('sig','radbas','tribas','relu','purelin'))
7  > model <- fit(tune, x = io_train$input,
8          y = io_train$output, ranges)
```

able at the International Fertilizer Association (IFA)[3] as a proof of concept of the present framework. This dataset of fertilizers was explored in deep in previous work [25]. It contains data on the annual consumption of three fertilizers (K_2O, N, P_2O_5) among the top ten main consumer countries. Each time series contains 60 observations from 1961 to 2020. Observations from 1961–2016 are used for training, and observations from 2017–2020 are used for testing. For the evaluation, we selected Brazil, the third major fertilizer consumer. All coding for the experimental evaluation is available[4].

The goal of this paper is to explore multiple facets of TSPredIT. The first experiment evaluated the effect of data transformation (`swminmax`, `diff`, `an`, `gminmax`) during prediction using `MLP` as `MLM`. Figure 3 compares these methods during testing for the three fertilizers K_2O, N, and P_2O_5 in Brazil. It presents the `sMAPE` error during testing. The `swminmax` outperformed other methods for K_2O and N. However, in P_2O_5, `an` was better followed close by `swminmax`.

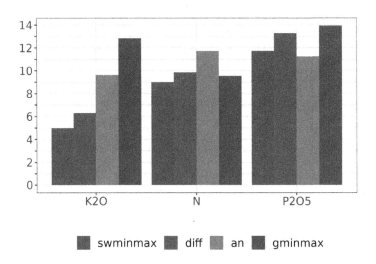

Fig. 3. Comparison of data transformations applied (`swminmax`, `diff`, `an`, `gminmax`)

A second evaluation explored the adoption of data augmentation techniques while fixing both `MLP` as `MLM` and `swminmax` as a data preprocessing technique. Figure 4 compares the performance of not applying data augmentation (none), using jittering (jitter) and warping stretching (stretch). As it can be observed, none was better both in K_2O and P_2O_5. However, for N, both jitter and stretch were worth value. The prediction performance increased by more than 1%. The data augmentation technique was seamlessly applied during time series cross-validation for training during hyperparameter optimization.

---

[3] http://www.fertilizer.org.
[4] https://eic.cefet-rj.br/~dal/tspredit/.

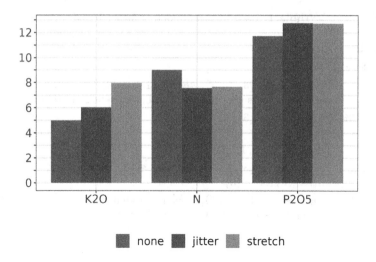

Fig. 4. Comparison of data augmentation applied (**none, jitter, stretch**)

Finally, the third evaluation studied the adoption of different MLM for predicting P_2O_5 using an. Figure 5 presents the evaluation of using Conv1D, ELM, MLP, RFR, SVM, LSTM for this scenario. Table 1 presents the hyperparameters explored.

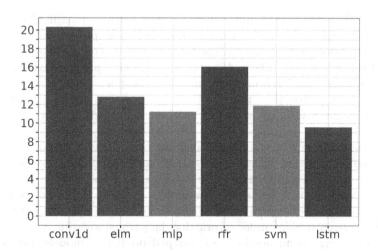

Fig. 5. Comparison of evaluation of using different MLM for predicting P_2O_5 using an

All these methods, except for Conv1D and LSTM, explored a similar amount of hyperparameter combinations (about 250 options each). The methods mostly could not improve the performance of the MLP. The exception was LSTM, which improved prediction by more than 1%.

Table 1. Used hyperparameters for each MLM

MLM	Hyperparameters
Conv1D	$input\_size \in \{3, \ldots, 7\}$
ELM	$input\_size \in \{3, \ldots, 7\}$, $nhid \in \{1, \ldots, 20\}$, $actfun \in \{$ "sig", "radbas", "tribas", "relu", "purelin"$\}$
LSTM	$input\_size \in \{3, \ldots, 7\}$
MLP	$input\_size \in \{3, \ldots, 7\}$, $size \in \{1, \ldots, 10\}$, $decay \in seq(0, 1, 1/9)$
RFR	$input\_size \in \{3, \ldots, 7\}$, $nodesize \in \{5, \ldots, 10\}$, $ntree \in \{1, \ldots, 10\}$
SVM	$input\_size \in \{3, \ldots, 7\}$, $kernel =$ "radial", $epsilon = seq(0, 1, 0.1)$, $cost = seq(20, 100, 20)$

As a proof of concept, these results can explore the capability of TSPredIT in combing a broad range of data preprocessing techniques and state-of-the-art MLM. Besides, the framework can provide hyperparameter optimization exploring both features, aiding the selection choice for these methods.

7   Conclusions

This paper presented TSPredIT, which extends features presented in TSPred [31]. It automates the entire time series prediction process by supporting hyperparameter optimization that combines time series data preprocessing and machine learning tuning. The architecture of TSPredIT provides five main modules. It includes data preprocessing, modeling support, prediction evaluation, and model tuning. Together, they are used to support the time series prediction process. It is made available as an extended version of DAL Toolbox Package at GitHub[5].

The combination of time series transformation methods in prediction with decomposed time series, transformation and model parameter selection, multi-step or one-step-ahead prediction, rolling origin evaluation, and the management of sliding windows is a key differentiation for TSPredIT. Several benchmark datasets from time series prediction competitions come bundled with TSPredIT. This new version enables users to practice data transformation and prediction methods, gaining confidence in the developed prediction models. Besides, the framework was designed to enable users to implement their customized methods. For example, both LSTM and Conv1D were added to TSPredIT as customized methods implemented in Python. Future updates will expand the range of implemented preprocessing methods, MLM, and evaluation metrics, especially empowering the hyperparameter selection targeting choosing combinations of data preprocessing and MLM that led to more stable models. This work leaves room for

---

[5] https://cefet-rj-dal.github.io/tspredit/.

future implementation including multivariate time series, cost of computation time, and short and long term prediction in classical problems.

Acknowledgements. The authors thank CNPq, CAPES (finance code 001), and FAPERJ for partially sponsoring this research.

References

1. Bergstra, J., Bengio, Y.: Random search for hyper-parameter optimization. J. Mach. Learn. Res. **13**, 281–305 (2012)
2. Bischl, B., et al.: mlr: machine learning in R. J. Mach. Learn. Res. **17**(170), 1–5 (2016)
3. Box, G.E.P., Jenkins, G.M., Reinsel, G.C., Ljung, G.M.: Time Series Analysis: Forecasting and Control. Wiley, Hoboken (2015)
4. Cheng, C., et al.: Time series forecasting for nonlinear and non-stationary processes: a review and comparative study. IIE Trans. (Ins. Ind. Eng.) **47**(10), 1053–1071 (2015). https://doi.org/10.1080/0740817X.2014.999180
5. Davydenko, A., Fildes, R.: Measuring forecasting accuracy: the case of judgmental adjustments To SKU-level demand forecasts. Int. J. Forecast. **29**(3), 510–522 (2013). https://doi.org/10.1016/j.ijforecast.2012.09.002
6. Diebold, F., Lopez, J.: 8 Forecast evaluation and combination. Handb. Stat. **14**, 241–268 (1996). https://doi.org/10.1016/S0169-7161(96)14010-4
7. Diebold, F., Mariano, R.: Comparing predictive accuracy. J. Bus. Econ. Stat. **20**(1), 134–144 (2002). https://doi.org/10.1198/073500102753410444
8. Eugster, M.J.A., Leisch, F.: Bench plot and mixed effects models: first steps toward a comprehensive benchmark analysis toolbox. In: Brito, P. (ed.) Compstat 2008, pp. 299–306. Physica Verlag, Heidelberg, Germany (2008)
9. Garcia, S., Luengo, J., Herrera, F.: Data Preprocessing in Data Mining. Springer (aug 2014). https://doi.org/10.1007/978-3-319-10247-4
10. Gujarati, D.N.: Essentials of Econometrics. SAGE (sep 2021)
11. Hao, J., Ho, T.: Machine learning made easy: a review of Scikit-learn package in python programming language. J. Educ. Behav. Stat. **44**(3), 348–361 (2019). https://doi.org/10.3102/1076998619832248
12. Hyndman, R., Khandakar, Y.: Automatic time series forecasting: The forecast package for R. J. Stat. Softw. **27**(3), 1–22 (2008). https://doi.org/10.18637/jss.v027.i03
13. Hyndman, R., Koehler, A., Snyder, R., Grose, S.: A state space framework for automatic forecasting using exponential smoothing methods. Int. J. Forecast. **18**(3), 439–454 (2002). https://doi.org/10.1016/S0169-2070(01)00110-8
14. Hyndman, R.J., Athanasopoulos, G.: Forecasting: principles and practice. OTexts (may 2018)
15. Izaú, L., et al.: Towards robust cluster-based hyperparameter optimization. In: Anais do Simpósio Brasileiro de Banco de Dados (SBBD), pp. 439–444. SBC (sep 2022). https://doi.org/10.5753/sbbd.2022.224330
16. Khalid, R., Javaid, N.: A survey on hyperparameters optimization algorithms of forecasting models in smart grid. Sustain. Cities Soc. **61**, 102275 (2020). https://doi.org/10.1016/j.scs.2020.102275
17. Kumar, A., McCann, R., Naughton, J., Patel, J.M.: Model selection management systems: the next frontier of advanced analytics. ACM SIGMOD Rec. **44**(4), 17–22 (2016). https://doi.org/10.1145/2935694.2935698

18. Lessmann, S., Baesens, B., Mues, C., Pietsch, S.: Benchmarking classification models for software defect prediction: a proposed framework and novel findings. IEEE Trans. Softw. Eng. **34**(4), 485–496 (2008). https://doi.org/10.1109/TSE.2008.35
19. Lim, B., Zohren, S.: Time-series forecasting with deep learning: a survey. Philos. Trans. R. Soc. A: Math. Phys. Eng. Sci. **379**(2194), 20200209 (2021). https://doi.org/10.1098/rsta.2020.0209
20. Lindemann, B., Müller, T., Vietz, H., Jazdi, N., Weyrich, M.: A survey on long short-term memory networks for time series prediction. In: Procedia CIRP. vol. 99, pp. 650–655 (2021). https://doi.org/10.1016/j.procir.2021.03.088
21. Lu, J., Liu, A., Dong, F., Gu, F., Gama, J., Zhang, G.: Learning under concept drift: a review. IEEE Trans. Knowl. Data Eng. **31**(12), 2346–2363 (2019). https://doi.org/10.1109/TKDE.2018.2876857
22. Moreno, A.V., Rivas, A.J.R., Godoy, M.D.P.: predtoolsTS: Time Series Prediction Tools. Tech. rep.,https://cran.r-project.org/package=predtoolsTS (2018)
23. Mumuni, A., Mumuni, F.: Data augmentation: a comprehensive survey of modern approaches. Array **16**, 100258 (2022). https://doi.org/10.1016/j.array.2022.100258
24. Ogasawara, E., Martinez, L., De Oliveira, D., Zimbrão, G., Pappa, G., Mattoso, M.: Adaptive normalization: a novel data normalization approach for non-stationary time series. In: Proceedings of the International Joint Conference on Neural Networks (2010). https://doi.org/10.1109/IJCNN.2010.5596746
25. Pacheco, C., et al.: Exploring data preprocessing and machine learning methods for forecasting worldwide fertilizers consumption. In: Proceedings of the International Joint Conference on Neural Networks. vol. 2022-July (2022). https://doi.org/10.1109/IJCNN55064.2022.9892325
26. Pedregosa, F., et al.: Scikit-learn: machine learning in Python. J. Mach. Learn. Res. **12**, 2825–2830 (2011)
27. Ramey, J.A.: sorting hat: sorting hat. Tech. rep., https://cran.r-project.org/web/packages/sortinghat/index.html (2013)
28. Ran, Z.Y., Hu, B.G.: Parameter identifiability in statistical machine learning: a review. Neural Comput. **29**(5), 1151–1203 (2017). https://doi.org/10.1162/NECOa00947
29. Salles, R., Assis, L., Guedes, G., Bezerra, E., Porto, F., Ogasawara, E.: A framework for benchmarking machine learning methods using linear models for univariate time series prediction. In: Proceedings of the International Joint Conference on Neural Networks. vol. 2017-May, pp. 2338–2345 (2017). https://doi.org/10.1109/IJCNN.2017.7966139
30. Salles, R., Belloze, K., Porto, F., Gonzalez, P., Ogasawara, E.: Nonstationary time series transformation methods: an experimental review. Knowl.-Based Syst. **164**, 274–291 (2019). https://doi.org/10.1016/j.knosys.2018.10.041
31. Salles, R., Pacitti, E., Bezerra, E., Porto, F., Ogasawara, E.: TSPred: a framework for nonstationary time series prediction. Neurocomputing **467**, 197–202 (2022). https://doi.org/10.1016/j.neucom.2021.09.067
32. Sarwar Murshed, M., Murphy, C., Hou, D., Khan, N., Ananthanarayanan, G., Hussain, F.: Machine learning at the network edge: a survey. ACM Comput. Surv. **54**(8), 1–37 (2022). https://doi.org/10.1145/3469029
33. Talavera, E., Iglesias, G., González-Prieto, A., Mozo, A., Gómez-Canaval, S.: Data Augmentation techniques in time series domain: A survey and taxonomy (jun 2022). https://doi.org/10.48550/arXiv.2206.13508,http://arxiv.org/abs/2206.13508
34. Von Luxburg, U.: A tutorial on spectral clustering. Stat. Comput. **17**(4), 395–416 (2007). https://doi.org/10.1007/s11222-007-9033-z

35. Wen, Q., et al.: Time series data augmentation for deep learning: a survey. In: IJCAI International Joint Conference on Artificial Intelligence, pp. 4653–4660 (2021)
36. Wickham, H.: Advanced R. CRC Press, second edn. (may 2019)

A Guide to the Tucker Tensor Decomposition for Data Mining: Exploratory Analysis, Clustering and Classification

Annabelle Gillet[1]([✉]), Éric Leclercq[1], and Lucile Sautot[2]

[1] LIB Univ. Bourgogne Franche Comté EA7534, Dijon, France
{annabelle.gillet,eric.leclercq}@u-bourgogne.fr
[2] UMR TETIS, AgroParisTech, Montpellier, France
lucile.sautot@agroparistech.fr

Abstract. Tensors are powerful multi-dimensional mathematical objects, that easily embed various data models such as relational, graph or time series. Furthermore, tensor decomposition operators are of great utility to reveal hidden patterns and complex relationships in data. Among these decompositions, the Tucker decomposition allows to factorize a tensor into a smaller core tensor and a set of factor matrices. In this article, we propose to study the capabilities of the Tucker decomposition when it is used in data mining techniques such as exploratory analysis, clustering and classification of data. We apply these different techniques on practical examples using several datasets having a ground truth. It is a preliminary work to add the Tucker decomposition to the Tensor Data Model, a model aiming at making tensors data-centric, and at optimizing operators in order to enable the manipulation of large tensors.

Keywords: Data mining · Tensor decomposition · Tucker decomposition

1 Introduction

When facing the volume and the variety of data, data mining techniques are often used to extract value. These techniques are rather diverse, and can consist in, for example, finding patterns in data, clustering similar elements, or training a model in order to classify new data [32]. However, depending on the technique used, data often have to be transformed in order to fit the data model required by the algorithm. When doing so, if the data model used is too restrictive to fully represent the data, the result obtained can be of a lesser quality than one obtained with a data model that allows to fully represent the characteristics of data.

In this context, tensors are a valuable solution [7]. Indeed, their multi-dimensional nature allows to easily embed different data models. For example, a tensor

© Springer-Verlag GmbH Germany, part of Springer Nature 2023
A. Hameurlain et al. (Eds.): *TLDKS LIV*, LNCS 14160, pp. 56–88, 2023.
https://doi.org/10.1007/978-3-662-68014-8_3

can naturally contains the adjacency matrix of a graph, but also more complex representation of graphs such as the labelled one [3]. Time series can also be represented along with their context, thus allowing to give more insights regarding data. Furthermore, tensors have powerful analytical operators, the tensor decompositions, that are used for various purposes such as dimensionality reduction, noise elimination, identification of latent factors, pattern discovery, ranking, recommendation or data completion. They are applied in a wide range of applications, including genomics [18], analysis of health records [46], graph mining [44] and identification and evolution of communities in social networks [3,34]. Papalexakis et al. in [35] review major usages of tensor decompositions in data mining applications.

One of the decompositions is the Tucker decomposition, that factorizes a tensor with N dimensions into a smaller core tensor and a set of N factor matrices, i.e., one for each dimension. The columns of the factor matrices can be seen as the features for the concerned dimension, and the lines as the signature of a specific element of the dimension over the features. The core tensor is also of major importance, as it represents the relationships of the features among dimensions. However, these different elements make the results of the Tucker decompositon tricky to interpret, especially compared to more straightforward decompositions, such as the CANDECOMP/PARAFAC [41] that does not produce a core tensor.

In this article, we study how different data mining techniques can be applied with the Tucker decomposition. These techniques include the exploratory analysis, that aims at finding patterns in data without having particular knowledge regarding the specificities of the data, the clustering, that gathers similar elements without supervision, and the classification, that gives a class to a new element depending on a model trained on known data. Several datasets covering different domains have been used to illustrate these techniques. It is a preliminary work aiming at integrating the Tucker decomposition into the Tensor Data Model [15, 28]. TDM adds the notion of schema and data manipulation operators to tensors, in order to make them data-centric and to avoid technical and functional errors brought by the manipulation of dimensions and elements of dimensions solely through integer indexes [39]. It also uses optimization techniques to allow the execution of operators, including the decompositions, on large-scale data [13].

The remaining of this article is organized as follows: Sect. 2 gives an overview of tensors and of some main operators, Sect. 3 details how main data models can be embedded into tensors, Sect. 4 presents the Tucker decomposition and two major algorithms to compute it, Sect. 5 relates of data mining techniques available with the Tucker decomposition that have been experimented on datasets, Sect. 6 evaluates the robustness of the Tucker decomposition regarding missing values, and finally Sect. 7 concludes the article and presents perspectives of future works.

2 Background of Tensors

Tensors are general abstract mathematical objects which can be considered according to various points of view, such as a multi-linear application or as the generalization of matrices to multiple dimensions. We will use the definition of a tensor as an element of the set of the functions from the product of N sets $I_j, j = 1, \ldots, N$ to $\mathbb{R} : \boldsymbol{\mathcal{X}} \in \mathbb{R}^{I_1 \times I_2 \times \cdots \times I_N}$, where N is the number of dimensions of the tensor or its order or its mode (see Fig. 1). Table 1 summarizes the notations used in this article. We adopt the same notation as Cichocki in [7].

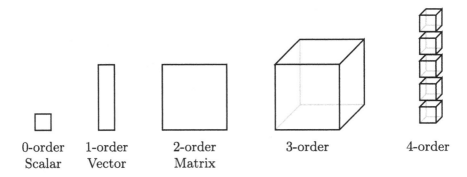

| 0-order | 1-order | 2-order | 3-order | 4-order |
| Scalar | Vector | Matrix | | |

Fig. 1. Tensors of different orders

Tensor operators, by analogy with operations on matrices and vectors, are multiplications, transpositions, matricizations (or unfolding) and decompositions (also named factorizations). We only highlight the most significant operators on tensors and matrices which are used in Tucker decomposition algorithms. The reader can consult [7,27] for an overview of the major operators.

Table 1. Symbols and operators used

Symbol	Definition
$\boldsymbol{\mathcal{X}}$	A tensor
$\mathbf{X}_{(n)}$	Matricization of a tensor $\boldsymbol{\mathcal{X}}$ on mode-n
a	A scalar
$\mathbf{v}$	A column vector
$\mathbf{M}$	A matrix
$\circ$	Outer product
$\otimes$	Kronecker product
$\mathbf{A}^{\otimes -n}$	$\mathbf{A}^{(N)} \otimes \cdots \otimes \mathbf{A}^{(n+1)} \otimes \mathbf{A}^{(n-1)} \cdots \otimes \mathbf{A}^{(1)}$
$[\mathbf{M}]_+$	Replace negative elements by 0 or small positive value
$\times_n$	Mode-n product
$\boldsymbol{\mathcal{X}} \times_{-n} \{\mathbf{A}\}$	$\boldsymbol{\mathcal{X}} \times_1 \mathbf{A}^{(1)} \cdots \times_{n-1} \mathbf{A}^{(n-1)} \times_{n+1} \mathbf{A}^{(n+1)} \cdots \times_N \mathbf{A}^{(N)}$
$\|\boldsymbol{\mathcal{X}}\|_F$	Frobenius norm

A **fiber** noted $\boldsymbol{\mathcal{Y}}_{i_1,\ldots,i_{n-1},:,i_{n+1},\ldots,i_N}$ consists in extracting a vector $\mathbf{v} \in \mathbb{R}^{I_n}$ from the dimension n of a tensor $\boldsymbol{\mathcal{Y}} \in \mathbb{R}^{I_1 \times I_2 \times \cdots \times I_n \times \cdots \times I_N}$. To do so, all the dimensions except the one to extract are fixed on a specific index, and the values of the vector are obtained with:

$$v_{i_n} = y_{i_1, i_2 \ldots, i_{n-1}, i_n, i_{n+1}, \ldots, i_N}$$

Slices are close to fibers, and aim at extracting a matrix $\mathbf{M} \in \mathbb{R}^{I_{n_1} \times I_{n_2}}$ from the dimensions n_1 and n_2 of a tensor $\mathcal{Y} \in \mathbb{R}^{I_1 \times I_2 \times \cdots \times I_{n_1} \times \cdots \times I_{n_2} \times \cdots \times I_N}$. All the dimensions except two are fixed on a specific index, and the values are obtained with:

$$m_{i_{n_1}, i_{n_2}} = y_{i_1, i_2, \ldots, i_{n_1}-1, i_{n_1}, i_{n_1}+1, \ldots, i_{n_2}-1, i_{n_2}, i_{n_2}+1, \ldots, i_N}$$

The concept of fibers and slices can be extended to extract a n-order sub-tensor from a N-order tensor with $n < N$, by fixing all the dimensions on a specific index except for n dimensions.

The **outer product** between a tensor $\mathcal{Y} \in \mathbb{R}^{I_1 \times I_2 \times \cdots \times I_N}$ and another tensor $\mathcal{X} \in \mathbb{R}^{J_1 \times J_2 \times \cdots \times J_M}$ noted $\mathcal{Y} \circ \mathcal{X}$ produces a tensor $\mathcal{Z} \in \mathbb{R}^{I_1 \times I_2 \times \cdots \times I_N \times J_1 \times J_2 \times \cdots \times J_M}$ in which the elements are equal to:

$$z_{i_1, i_2, \ldots, i_N, j_1, j_2, \ldots, j_M} = y_{i_1, i_2, \ldots, i_N} x_{j_1, j_2, \ldots, j_M}$$

It allows to combine all the values from both tensors, by having as many dimensions as the sum of the order of the input tensors. For example, when applying the outer product on two vectors (1-order tensors), it will produce a matrix (2-order tensor), in which an element $e_{i,j}$ corresponds to the ith element of the first vector multiplied by the jth element of the second vector.

The **mode-n product** allows to multiply a tensor by a matrix or a vector. For a tensor $\mathcal{X} \in \mathbb{R}^{J_1 \times J_2 \times \cdots \times J_n \times \cdots \times J_N}$ and a matrix $\mathbf{M} \in \mathbb{R}^{I_n \times J_n}$, the result of the mode-n product noted $\mathcal{X} \times_n \mathbf{M}$ is a new tensor $\mathcal{Y} \in \mathbb{R}^{J_1 \times J_2 \times \cdots \times I_n \times \cdots \times J_N}$ where:

$$y_{j_1, \ldots, j_{n-1}, i_n, j_{n+1}, \ldots, j_N} = \sum_{j_n=1}^{J_n} x_{j_1, \ldots, j_{n-1}, j_n, j_{n+1}, \ldots, j_N} m_{i_n, j_n}$$

It modifies the size of the dimension n from J_n to I_n. It can be compared to a standard matrix multiplication: for all the indexes of the dimension n, a fiber $\mathbf{v_1} \in \mathbb{R}^{J_n}$ is obtained, and the multiplication $\mathbf{M} \times \mathbf{v_1}$ is performed, resulting in a vector $\mathbf{v_2} \in \mathbb{R}^{I_n}$ that replaces the fiber extracted from the tensor.

The mode-n product between a N-order tensor $\mathcal{X} \in \mathbb{R}^{I_1 \times \cdots \times I_{n-1} \times I_n \times I_{n+1} \times \cdots \times I_N}$ and a vector $\mathbf{v} \in \mathbb{R}^{I_n}$ noted $\mathcal{X} \times_n \mathbf{v}$ produces a $(N-1)$-order tensor $\mathcal{Y} \in \mathbb{R}^{I_1 \times \cdots \times I_{n-1} \times I_{n+1} \times \cdots \times I_N}$ where:

$$y_{i_1, \ldots, i_{n-1}, i_{n+1}, \ldots, i_N} = \sum_{i_n=1}^{I_n} x_{i_1, \ldots, i_{n-1}, i_n, i_{n+1}, \ldots, i_N} v_{i_n}$$

The behavior of the mode-n product between a tensor and a vector is the same as the one between a tensor and a matrix, except that the product is a dot product between $\mathbf{v}$ and $\mathbf{v_1} \in \mathbb{R}^{I_n}$, that yields a scalar. Thus, the resulting size of the dimension n is 1, and it can be removed from the tensor.

The **mode-n matricization** of a tensor $\mathcal{X} \in \mathbb{R}^{I_1 \times I_2 \times \cdots \times I_N}$ noted $\mathcal{X}_{(n)}$ produces a matrix $\mathbf{M} \in \mathbb{R}^{I_n \times \Pi_{j \neq n} I_j}$, where:

$$m_{i_n,j} = x_{i_1,\ldots,i_n,\ldots,i_N} \text{ with } j = 1 + \sum_{\substack{k=1 \\ k \neq n}}^{N} (i_k - 1) \prod_{\substack{m=1 \\ m \neq n}}^{k-1} I_m$$

The matricization is a useful operator, that allows to convert a tensor into a matrix without losing information, in order to apply matrix operators such as the Hadammard product, the Kronecker product or the Khatri-Rao product [7].

The **Kronecker product** between two matrices $\mathbf{A} \in \mathbb{R}^{I \times J}$ and $\mathbf{B} \in \mathbb{R}^{K \times L}$ noted $\mathbf{A} \otimes \mathbf{B}$ produces a matrix $\mathbf{C} \in \mathbb{R}^{(IK) \times (JL)}$, in which every elements of $\mathbf{A}$ are multiplied by the matrix $\mathbf{B}$:

$$c_{m,n} = a_{i,j} b_{k,l} \text{ where } m = k + (i - 1)K \text{ and } n = l + (j - 1)L$$

The **Frobenius norm** of a tensor $\mathcal{X} \in \mathbb{R}^{I_1 \times I_2 \times \cdots \times I_N}$ noted $\|\mathcal{X}\|_F$ is computed with:

$$\|\mathcal{X}\|_F = \sqrt{\sum_{i_1=1}^{I_1} \cdots \sum_{i_N=1}^{I_N} |x_{i_1,\ldots,i_N}|^2}$$

It is often used on the difference between two tensors in order to estimate their similarity.

3   From Data Models to Tensors

By means of their multi-dimensional nature, tensors can represent various data models. This section highlights how common data models can be embedded into a tensor.

3.1   Key-Value Model

The key-value model stores data as $(key, value)$ pairs [4]. Thus, 1-order tensors $\mathcal{KV} \in \mathbb{R}^{|key|}$ can be used to represent this model, with the dimension storing the keys, and the values of the tensor being the value of the pairs. With tensors, the key-value model can be extended to a multi-keys model, in which a value is indexed by several keys. To do so, the tensor must have as many dimensions as the number of keys in a pair, i.e., for K keys the tensor will have K dimensions $\mathcal{MKV} \in \mathbb{R}^{|key_1| \times \cdots \times |key_K|}$.

3.2   Relational and Column Models

A relation R (or table) [24] is a set of tuples $(v_1, v_2, \ldots, v_N)$, where each element v_i is a member of a domain Dom_i, so the set-theoretic relation R is a subset of the cartesian product of the domain $Dom_1 \times \cdots \times Dom_N$. With this definition, any relation can be represented with a N-order tensor $\mathcal{R} \in \mathbb{R}^{|Dom_1| \times \cdots \times |Dom_N|}$ in which the values are the number of occurrences of this combination of elements.

OLAP data cubes [16], that are obtained from GROUP BY queries, are naturally embedded into tensors because they already represent a multi-dimensional array (see Fig. 2). By formally defining a data cube with $f : (A_1, \ldots, A_N) \rightarrow v$, we can use a N-order tensor $\mathcal{DC} \in \mathbb{R}^{|Dom_{A_1}| \times \cdots \times |Dom_{A_N}|}$ populated with the v values.

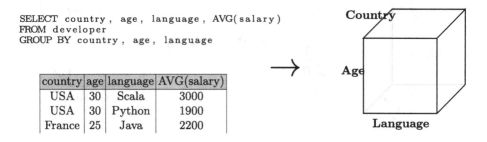

Fig. 2. Building a tensor from a GROUP BY query

This representation can also model a relation in which the association of $N - 1$ elements guarantees a unique combination (e.g., a primary key), and a last element that carries a specific value. For example, for tuples that represent a user, its city and its age, the users and the cities can each be embedded into a dimension, and the age can be the value of the tensor.

The column model, which we consider as a semi-structured model with a fixed schema (i.e., that has a fixed number of columns), includes CSV files and dataframes [36]. This type of model is close to the relational one, thus the same mechanisms of modelling can be used.

3.3 Graph Models

A simple graph $G = (V, E)$, where V is the set of the vertices (or nodes) and $E \subseteq V \times V$ the set of the edges (or links), can be represented by its adjacency matrix, and thus by a 2-order tensor $\mathcal{G} \in \mathbb{R}^{|V| \times |V|}$ that can take into consideration the direction and the weight of the edges.

However, this is a basic representation, and embedding a graph into a tensor can be even more useful. In a lot of real world graphs, the edges are labelled. An edge labelled graph has its edges defined by $E \subseteq V \times Lab \times V$, where Lab is a set of labels [2]. For example, in a social network the interactions among users are represented by edges that can carry more information, such as the time of the interaction or important words (or hashtags) used in the message. Tensors, as opposed to classical graph representations, can naturally put each type of label into a dimension and have a more complete representation of the data, i.e., for a graph with L different labels and $|Lab_i|$ the number of distinct values taken by the label Lab_i, we have $\mathcal{LG} \in \mathbb{R}^{|V| \times |V| \times |Lab_1| \times \cdots \times |Lab_L|}$ (see Fig. 3). The previous example can be modeled as a 4-order tensor, with one dimension for the source user, one for the destination user, one for the hashtag and one for the time.

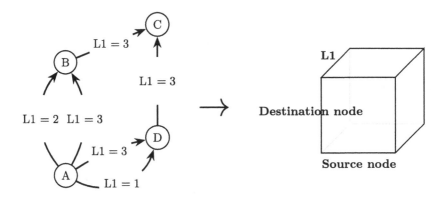

Fig. 3. Building a tensor from a labelled graph

Multi-layer networks [26] are also of high interest. A multi-layer network $M = (V_M, E_M, V, \mathbf{L})$ with D layers has a set of vertices V and a set of layers $\mathbf{L} = \{L_1, \ldots, L_D\}$. A vertex can be present in multiple layers, so $V_M \subseteq V \times L_1 \times \cdots \times L_D$. Thus, the edges link a vertex within a layer to another vertex within a layer, that is $E_M \subseteq V_M \times V_M$. Each layer represents a category of vertices. This type of graph can be embedded into a 4-order tensor, i.e., $\mathcal{M} \in \mathbb{R}^{|V| \times |V| \times |L| \times |L|}$, with one dimension for the source vertex, one for the destination vertex, one for the source layer and one for the destination layer. On top of that, by adding dimensions to this representation, tensors can represent easily labelled multi-layer networks.

3.4   Time Series

A time series $Y = (Y_t : t \in T)$ follows the evolution of a metrics for a given element during time [17]. A 1-order tensor $\mathcal{Y} \in \mathbb{R}^{|T|}$ can represent a standard time series by storing the time in the dimension. However, tensors can shine as model for time series, as they allow to integrate much more parameters of the creation of the time series (e.g., the sensor, the location), each parameter being represented as a dimension (see Fig. 4). By doing so, time series are viewed in their global context, and therefore it adds more precision and information to the analyses performed on them.

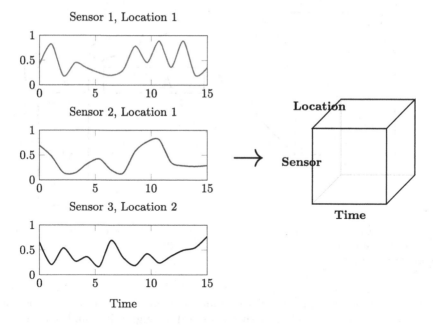

Fig. 4. Building a tensor from time series

3.5   Images and Videos

A gray scale image of size $x \times y$ is a matrix $\mathbf{GI} \in \mathbb{R}^{x \times y}$, thus a 2-order tensor $\boldsymbol{\mathcal{GI}} \in \mathbb{R}^{x \times y}$. More complex images that use multiple color channels (e.g., RGB, YUV, CYMK) can be embedded in a 3-order tensor $\boldsymbol{\mathcal{CI}} \in \mathbb{R}^{x \times y \times c}$, where c is the number of channels. Videos can be considered as a succession of images, called frames. In this configuration, a video is a 4-order tensor $\boldsymbol{\mathcal{V}} \in \mathbb{R}^{x \times y \times c \times f}$ where f is the number of frames.

4   Tucker Decomposition

The Tucker decomposition [45] factorizes a N-order tensor $\boldsymbol{\mathcal{X}} \in \mathbb{R}^{I_1 \times \cdots \times I_N}$ into a core tensor $\boldsymbol{\mathcal{G}} \in \mathbb{R}^{R_1 \times \cdots \times R_N}$ and N factor matrices $\mathbf{A}^{(n)} \in \mathbb{R}^{I_n \times R_n}$. The Fig. 5 shows a representation of the Tucker decomposition applied on a 3-order tensor. The ranks $R_1, \ldots, R_N$ are input parameters that determine the number of column vectors (that can be seen as the different features) for each factor matrix. For each rank R_i, we have $R_i \leq I_i$, and most of the time $R_i \ll I_i$. The input tensor can be approximated from the result of the decomposition with:

$$\boldsymbol{\mathcal{X}} \simeq \boldsymbol{\mathcal{G}} \times_1 \mathbf{A}^{(1)} \cdots \times_N \mathbf{A}^{(N)}$$

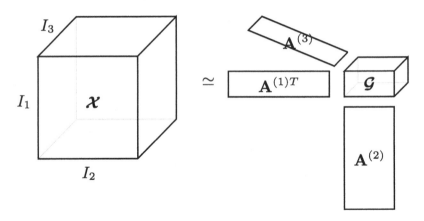

Fig. 5. The Tucker decomposition, with $\mathcal{X} \in \mathbb{R}^{I_1 \times I_2 \times I_3}$ the input tensor, $\mathcal{G} \in \mathbb{R}^{R_1 \times R_2 \times R_3}$ the core tensor, and $\mathbf{A}^{(1)} \in \mathbb{R}^{I_1 \times R_1}$, $\mathbf{A}^{(2)} \in \mathbb{R}^{I_2 \times R_2}$ and $\mathbf{A}^{(3)} \in \mathbb{R}^{I_3 \times R_3}$ the factor matrices

The order of the elements of the dimensions of the input tensor does not impact the result of the decomposition. Indeed, changing it would only reorder the line vectors of the factor matrices, as each line vector stores the result of the decomposition for a given element on the dimension corresponding to the factor matrix.

To compute the Tucker decomposition, several algorithms have been proposed. Each has some advantages, as for example imposing more easily the orthogonality constraint (that allows a good clustering of elements) or the non-negativity constraint (that provides more interpretable results). Two major algorithms are presented in this section: the Higher-Order Orthogonal Iteration (HOOI) and the Hierarchical Alternating Least Squares Non-negative Tucker Decomposition (HALS-NTD).

4.1   Higher-Order Orthogonal Iteration Algorithm

The HOOI algorithm [8] is the most famous one to compute the Tucker decomposition (Algorithm 1). It depends primarily on the Singular Value Decomposition (SVD), that it extends to cope with multiple dimensions.

The HOOI starts by initializing the factor matrices, by matricizing the original tensor on each dimension in order to apply the SVD and to use the R_n left singular vectors (matrix $\mathbf{U}$ of the result of the SVD truncated at the R_n^{th} column) as factor matrices. During the iterative phase (lines 2 to 7), each factor matrix is improved. To do so, a partial core tensor $\mathcal{Y} \in \mathbb{R}^{R_1 \times \cdots \times I_n \times \cdots \times R_N}$ is computed by performing the mode-n product on the original tensor and all the factor matrices except the one being improved. This partial core tensor is then matricized on the mode corresponding to the concerned dimension, and the SVD is executed on it. As for the initialization step, the R_n left singular vectors are used as the new factor matrix. The iterative phase allows to refine the result, as the partial core tensor takes into consideration the other factor matrices. So, each factor matrix is improved depending on the other factor matrices, thus reinforcing the discovering of relationships among elements of dimensions. When a convergence criteria is met, the final core tensor is computed from the original core tensor and all the factor matrices (line 8).

Algorithm 1. Higher-Order Orthogonal Iteration (HOOI)

Require: Tensor $\mathcal{X} \in \mathbb{R}^{I_1 \times I_2 \times \cdots \times I_N}$ and target ranks $R_1, \ldots, R_N$
Ensure: Core tensor $\mathcal{G} \in \mathbb{R}^{R_1 \times R_2 \times \cdots \times R_N}$ and factor matrices $\mathbf{U}^{(1)}, \ldots, \mathbf{U}^{(N)}$ with
$\quad \mathbf{U}^{(n)} \in \mathbb{R}^{I_n \times R_n}$
1: Initialize $\mathbf{U}^{(1)}, \ldots, \mathbf{U}^{(N)}$, with $\mathbf{U}^{(n)} \in \mathbb{R}^{I_n \times R_n}$, $\mathbf{U}^{(n)} \leftarrow SVD(\mathbf{X}_{(n)}).\mathbf{U}(:, 1 : R_n)$
2: **repeat**
3:    **for** $n = 1, \ldots, N$ **do**
4:      $\mathcal{Y} \leftarrow \mathcal{X} \times_N \mathbf{U}^{(N)T} \times_{n+1} \mathbf{U}^{(n+1)T} \times_{n-1} \mathbf{U}^{(n-1)T} \cdots \times_1 \mathbf{U}^{(1)T}$
5:      $\mathbf{U}_{(n)} \leftarrow SVD(\mathbf{Y}_{(n)}).\mathbf{U}(:, 1 : R_n)$
6:    **end for**
7: **until** $<$ convergence $>$
8: $\mathcal{G} \leftarrow \mathcal{X} \times_N \mathbf{U}^{(N)T} \cdots \times_1 \mathbf{U}^{(1)T}$

A simpler version of the HOOI algorithm exists, the Higher-Order Singular Value Decomposition (HOSVD), that removes the iterative part of the HOOI algorithm (lines 2 to 7). It is less precise, as the iterative part of the HOOI allows to refine the result until a convergence criterion is met.

The HOOI algorithm inherits from the orthogonality constraint of the SVD for the computation of the factor matrices. Thus, it works pretty well to cluster elements of a dimension depending on their behavior on the other dimensions. However, as the SVD produces matrices with positive and negative values, the HOOI is not well suited to impose the non-negativity constraint on factor matrices, as some negative values will be found (and must be removed) at each iteration.

An advantage of this algorithm is that it can easily be implemented on large tensors. The most costly operation is the computation of the SVD, that is found at the initialization (line 1) and during the iterative phase (line 5). During the iterations, as the SVD is executed on the mode-n matricized partial core tensor, that is relatively small compared to the matricized original tensor, the time and space complexity is reduced. At the initialization of the algorithm, it can be replaced with a random initialization to avoid the computation of the SVD on a too large matrix.

4.2 Hierarchical Alternating Least Squares Algorithm

The HALS-NTD algorithm [7] uses a different approach than the HOOI algorithm (Algorithm 2), even if the initialization step (line 1) can be done by using the HOSVD. An alternative version of the HALS-NTD was later proposed [37].

Algorithm 2. Hierarchical Alternating Least Squares (HALS-NTD)

Require: Tensor $\mathcal{X} \in \mathbb{R}^{I_1 \times I_2 \times \cdots \times I_N}$ and target ranks $R_1, \ldots, R_N$

Ensure: Core tensor $\mathcal{G} \in \mathbb{R}^{R_1 \times R_2 \times \cdots \times R_N}$ and factor matrices $\mathbf{A}^{(1)}, \ldots, \mathbf{A}^{(N)}$ with $\mathbf{A}^{(n)} \in \mathbb{R}^{I_n \times R_n}$

1: Initialize $\mathbf{A}^{(1)}, \ldots, \mathbf{A}^{(N)}$ with non-negativity constraint

2: $\mathcal{E} \leftarrow \mathcal{X} - \mathcal{G} \times_1 \mathbf{A}^{(1)} \cdots \times_N \mathbf{A}^{(N)}$

3: **repeat**

4:     **for** $n = 1, \ldots, N$ **do**

5:         **for** $r = 1, \ldots, R_n$ **do**

6:             $\mathbf{X}_{(n)}^{(r)} = \mathbf{E}_{(n)} + \mathbf{a}_r^{(n)} \left[\mathbf{G}_{(n)} \right]_r \mathbf{A}^{\otimes -n\,T}$

7:             $\mathbf{a}_r^{(n)} \leftarrow \left[\mathbf{X}_{(n)}^{(r)} \left[(\mathcal{G} \times_{-n} \{\mathbf{A}\})_{(n)} \right]_r^T \right]_+$

8:             $\mathbf{a}_r^{(n)} \leftarrow \mathbf{a}_r^{(n)} / \left\| \mathbf{a}_r^{(n)} \right\|_2$

9:             $\mathbf{E}_{(n)} \leftarrow \mathbf{X}_{(n)}^{(r)} - \mathbf{a}_r^{(n)} \left[\mathbf{G}_{(n)} \right]_r \mathbf{A}^{\otimes -n\,T}$

10:         **end for**

11:     **end for**

12:     **for** $r_1 = 1, \ldots, R_1, \ldots, r_N = 1, \ldots, R_N$ **do**

13:         $g_{r_1, \ldots, r_N} \leftarrow g_{r_1, \ldots, r_N} + \mathcal{E} \times_1 \mathbf{a}_{r_1}^{(1)} \cdots \times_N \mathbf{a}_{r_N}^{(N)}$

14:         $\mathcal{E} \leftarrow \mathcal{E} + \Delta_{g_{r_1, \ldots, r_N}} \mathbf{a}_{r_1}^{(1)} \circ \cdots \circ \mathbf{a}_{r_N}^{(N)}$

15:     **end for**

16: **until** < convergence >

The HALS-NTD starts also by initializing the factor matrices (line 1), but adds a non-negativity constraint to manipulate only positive values in the remaining of the algorithm. An error tensor $\mathcal{E} \in \mathbb{R}^{I_1 \times \cdots \times I_N}$ (noted $\mathbf{E}_{(n)}$ when it is matricized on dimension n), that stores the difference between the original tensor and the reconstructed tensor from the core tensor and the factor matrices, is computed (line 2). The iterative phase (lines 3 to 16) is more complex than the one of the HOOI algorithm. Rather than improving a whole factor matrix at a time, it improves a vector of a factor matrix at a time. To do so, at the line 6, the current vector (the one being improved) is put in relation with all the other factor matrices associated with the part of the core tensor representing the strength of the relationships of the current vector regarding the vectors of the other factor matrices. This result is added to the error tensor, and stored in $\mathbf{X}_{(n)}^{(r)}$, that can be seen as a matricized tensor representing the contribution of the current vector to the global result combined to the error tensor. At line 7, the current vector is improved by multiplying $\mathbf{X}_{(n)}^{(r)}$ with the part of the recon-structed tensor corresponding to the current vector. The current vector is then normalized with a L_2 norm (line 8), and the error tensor is updated (line 9). Once all the vectors of the factor matrices have been improved, the core tensor is updated from the previous core tensor, the error tensor and the new vectors of the factor matrices (line 13), and finally the error tensor is updated to integrate the changes in the core tensor (line 14).

The major advantage of the HALS-NTD is that it enforces the non-negativity constraint by imposing it during the initialization step, and then by improving the result without obtaining (and without having to eliminate) negative values during the iterative part (line 3 to 12). Thus, it eases the direct interpretation of the factor matrices as well as the core tensor.

However, as this algorithm computes the decomposition column vector by column vector for each factor matrices, it is computationally demanding, and harder to optimize than the HOOI one. Furthermore, there is almost no implementation of the HALS-NTD alogrithm. To the best of our knowledge, only Cichocki and Phan have provided a Matlab implementation in [7].

4.3   Related Work

The Tucker decomposition has been used in several kind of applications on specific data such as in social and collaboration network analysis, in web mining, in topic modelling, in recommendation systems, in urban computing, in vision, image or speech processing. While the number of works on the CANDECOMP/PARAFAC decomposition algorithm is significant, much less work has been done to study the Tucker decomposition algorithms. Some of these works focus on specific issues of algorithms to compute the Tucker decomposition.

In [22] the authors proposed the D-Tucker decomposition, as a fast and memory-efficient method for Tucker decomposition on large dense tensors using 3 phases: the approximation, the initialization, and the iterative phases. The main ideas is to compress an input tensor by computing randomized SVD of matrices sliced from the input tensor, and to obtain orthogonal factor matrices and a core tensor by using SVD results.

In [23] the authors proposed Tucker decomposition methods for large dense static tensors and online streaming data. They decomposed large dense tensors by using the randomized SVD, avoiding the reconstruction from SVD results, and carefully determining the order of operations.

Chachlakis et al. in [6] explored the use of L_1-norm for reformulation of the Tucker decomposition to overcome the effect of outliers. They also adapted two algorithms: the L_1-norm Higher-Order Singular Value Decomposition (L_1-HOSVD) and the L_1-norm Higher-Order Orthogonal Iterations (L_1-HOOI).

In [29], the authors defined a scalable GPU-based Tucker decomposition, which partitions large-scale tensors into subtensors to process them. They showed that their decomposition reduced the overhead on a single machine.

Most of the Tucker decomposition implementations make use of explicit matricizations and could introduce extra costs in terms of data conversion and memory usage. In [10] the authors proposed A-Tucker, a framework for input-adaptive and matricization-free Tucker decomposition of dense tensors. Their decomposition algorithm enables the switch of different solvers for the factor matrices and core tensor, and a machine-learning adaptive algorithm is applied to automatically cope with the variations of both the input data and the hardware. They showed that A-Tucker improves existing algorithm on GPUs.

Several applications of the Tucker decomposition can be found in the literature.

Fernandes et al. [12] presented an overview of tensor decompositions for analyzing time-evolving social networks and showed that while most of the approaches use the CANDECOMP/PARAFAC decomposition, Tucker is most appropriate for studying time evolving networks. Sun et al. [43] used Tucker on social network data in order to find clusters. They applied it on the Enron dataset. They also proposed visualization techniques based on graphs to display the result of the decomposition. Al-Sharoa et al. in [1] proposed an approach to determine sub-spaces across time which relies on Tucker decomposition.

Shao et al. in [40] developed a model for temporal knowledge graphs completion based on a specific tensor decomposition model for temporal knowledge graphs completion inspired by the Tucker decomposition, but only for 4-order tensors. For handling missing data, [30] introduced a Tucker decomposition with L_2 regularization and applied it on urban IoT data.

Romeo et al. [38] used the Tucker decomposition to cluster documents. Thanks to this decomposition, they were able to process documents in several languages in the same tensor, in order to find similarities in the whole dataset.

Huang et al. [20] compared the Tucker decomposition to the PCA and SVD associated to k-means. They ran experiments on three datasets of images to show the similarities among these algorithms. Zhou et al. [47] took a different approach and used the Tucker decomposition as a supervised learning algorithm. They obtained promising results to cluster images.

Cichocki in his book [7] showed several applications of various decompositions on small tensors, mainly for image and brain data signal analysis. In [19], the authors approximated both spectral and spatial information, and proposed a novel 3-order Tucker decomposition and a reconstruction detector for hyperspectral change detection. They designed a singular value energy accumulation method to determine the number of principal components in different factor matrices.

Brandoni et al. in [5] defined a method which can handle three or more order tensors in the Tensor-Train model and they proposed to tackle the memory consumption with a truncation strategy. For 3-order tensors, the Tensor-Train decomposition corresponds to the classical Tucker decomposition. They applied their method for image classification.

In [33] the authors used Tucker decomposition as the core of a deep neural network method for speech emotion recognition. 2D, 3D Spectrogram and Temporal Modulation Spectrogram are explored to investigate tensor factorization based architectures to capture salient information corresponding to emotion. Hidden layers are extracted from Tucker decomposition. The core tensor produced in each hidden layer is the feature associated with that factorization layer.

Due to the lack of implementation for the HALS-NTD algorithm, articles are mainly related to the HOOI algorithm. Thus, they only benefit from a part of the Tucker decomposition capabilities. They aim at clustering data, and do not rely on the direct interpretability of the factor matrices and the core tensor even if it brings valuable insights.

5 Data Mining Techniques

This section presents the datasets used for the experiments, as well as the different data mining techniques that can be applied with the Tucker decomposition. In [14], we observed that the HALS-NTD algorithm with its non-negativity constraint is best suited for producing interpretable results, while the HOOI algorithm with its orthogonality constraint is best suited for clustering tasks. So, in this article we focus only on the different data mining techniques without analyzing the impact of each algorithm on the techniques. The code of the experiments is available online[1] as well as the links to datasets in order to make the experiments reproducible.

5.1 Datasets Overview

To illustrate the different data mining techniques, we rely on several well-known datasets. They are rather diverse and concern different domain applications, such as image recognition, temporal graph or machine learning reference dataset.

Iris [2]

The Iris dataset contains characteristics of 150 flowers, namely the sepal width, the sepal length, the petal width and the petal length. There are 3 species of Iris flowers in this dataset, each being represented by 50 samples. This dataset is known for having one species that are linearly separable from the two others, and two species that are not linearly separable.

COIL-20 [31]

Fig. 6. The 20 objects of the COIL-20 dataset

---

[1] https://github.com/AnnabelleGillet/Tucker-experiments.
[2] https://archive.ics.uci.edu/ml/datasets/iris.

The COIL-20 dataset gathers 20 different objects (see Fig. 6). For each object, there are 72 pictures that represent the object in a specific position (a difference of 5° in the orientation of the object). The pictures have 128×128 pixels.

MNIST [9].

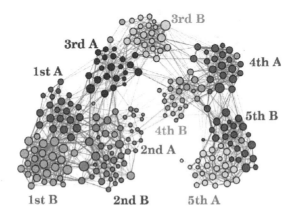

Fig. 7. An extract of the MNIST dataset

The MNIST dataset is composed of 70 000 images representing a hand-written digit, of 28×28 pixels (see Fig. 7). It is often used to evaluate algorithms of image classification, as the hand-written nature of the images induces a different complexity to deal with than a static object.

Primary School [42]

Fig. 8. Network overview of the primary school dataset

The primary school dataset represents the interactions among 232 students and 10 teachers in a French primary school, that contains 10 classes (see Fig. 8). The participants wore RFID devices, that recorded an interaction if it had lasted at least 20 seconds. The experiment was carried on during 2 days. The records are of the form (person1, person2, timestamp), and the class of each student is the ground truth of this dataset.

5.2   Exploratory Analysis

The Tucker decomposition can be used to highlight patterns of multi-dimensional data. Indeed, as each factor matrix gives information regarding elements of a dimension depending on their behavior on other dimensions, it helps to find structures or patterns in data. Furthermore, the core tensor allows to link this kind of information among all the dimensions, and thus to contextualise each insight.

To illustrate this use of the Tucker decomposition, we rely on the primary school dataset. We build a 3-order tensor of size $242 \times 242 \times 208$, with two symmetric dimensions used to represent the persons, and the third dimension to represent the time with a granularity of 5 min. If a person has been in contact with another person at a time t, then the value in the tensor indexed by the corresponding dimension values is 1.

This kind of use of the Tucker decomposition is best interpreted when the non-negativity constraint is enforced during the decomposition algorithm execution. So, in the experiment of this section, we use the HALS-NTD algorithm. We run the Tucker decomposition with ranks 13 (for the first person dimension), 13 (for the second person dimension) and 4 (for the time dimension). To choose these ranks, we ran the SVD for each dimension on the tensor matricized on the corresponding dimension, and we select as rank the number of significant singular values.

The factor matrix for the first dimension is shown in Fig. 9. Each line represents a rank, and the columns are the persons. The students are ordered by their class: the first columns are the students of the class 1A, then 1B, and so on until 5B, and the 10 teachers are the last 10 columns. We can distinguish 10 ranks in which each class appears distinctly, and three heterogeneous ranks.

Figure 10 shows the factor matrix for the time dimension. There are four distinct periods. The first period indicates an activity during class hours and the morning and afternoon breaks, the second period concerns the breaks, including the lunch one, the third period also covers the class hours with more activity at the end of the days, and the last one shows activity during morning and afternoon breaks and just before and after the lunch time.

For exploratory analysis, the role of the core tensor is also important: it gives insights regarding the strength of the relationships of the ranks among dimensions. For example, the $g_{1,1,1}$ value indicates if the vectors $\mathbf{a}_1^{(1)}$, $\mathbf{a}_1^{(2)}$ and $\mathbf{a}_1^{(3)}$ are strongly related or not.

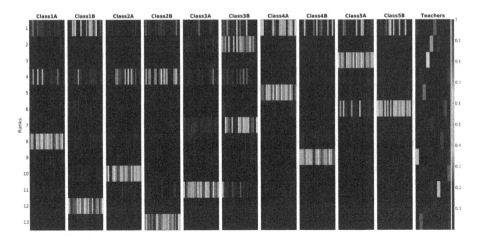

Fig. 9. Factor matrix for the dimension representing the persons in the primary school experiment

To illustrate the usefulness of the core tensor, we can focus on a particular rank of the first dimension and see how it is related to the ranks of the other dimensions. The Fig. 11 represents this mechanism when fixing the rank of the first dimension to the one corresponding to the class 1A.

This figure shows some interesting results. The first strongest value of the core tensor indicates that the class 1A has strong ties with itself, mainly at the break times and before and after the lunch break (Fig. 11a). It makes sense because at the breaks the students move from their classroom and go outside, so it creates more interactions among students. The second strongest value of the core tensor shows again a relationship of the class 1A with itself, but this time during the class hours (Fig. 11b). The third strongest value indicates a relationship between the class 1A and 1B during the breaks, including the lunch one (Fig. 11c). As the students of these two classes are of the same age, it is logical that they have more ties. Finally, the fourth value of the core tensor shows a relationship between the class 1A and a heterogeneous cluster that gathers students from grades 1, 2 and 3, during the breaks (Fig. 11d).

The Fig. 12 indicates the number of contacts that have occurred among classes, and it confirms the observations made from the result of the Tucker decomposition. In each class, most of the contacts are made with students of the same class, or with students with a close age.

To summarize, the advantages of using the Tucker decomposition for exploratory analysis are twofold: 1) the vectors of the factor matrices give insights regarding the elements that contribute to the rank; and 2) the core tensor allows to link the ranks of one factor matrix to the ranks of the other factor matrices, and thus it gives more context to the result, as for example in Fig. 11 where we have the temporal activity of the different communities.

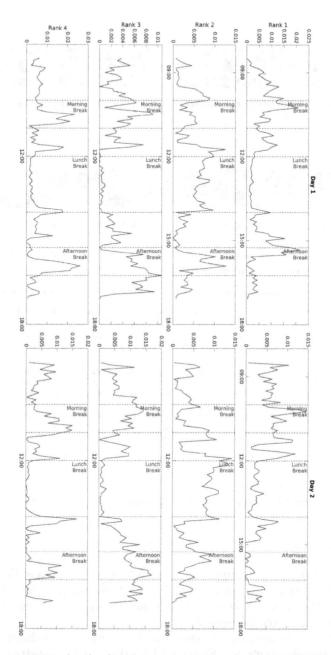

Fig. 10. Factor matrix for the dimension representing the time in the primary school experiment. The morning and afternoon breaks are approximate, as all the classes do not have the breaks at the same time

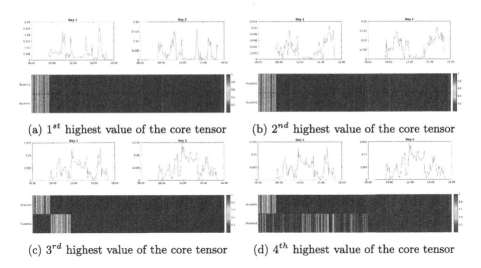

(a) 1st highest value of the core tensor   (b) 2nd highest value of the core tensor

(c) 3rd highest value of the core tensor   (d) 4th highest value of the core tensor

Fig. 11. Ranks from each factor matrix that are strongly related to each other when fixing the rank of the first dimension to the one representing the class 1A

	1st A	1st B	2nd A	2nd B	3rd A	3rd B	4th A	4th B	5th A	5th B	teachers
1st A	4505	1051	594	625	560	286	83	160	57	105	149
1st B	1051	9756	502	632	269	207	551	161	448	386	1084
2nd A	594	502	5401	1583	657	360	77	56	76	30	586
2nd B	625	632	1583	6270	712	373	119	36	41	54	508
3rd A	560	269	657	712	5537	2076	77	163	109	82	414
3rd B	286	207	360	373	2076	5926	248	193	154	219	282
4th A	83	551	77	119	77	248	4496	828	351	745	382
4th B	160	161	56	36	163	193	828	2843	119	346	168
5th A	57	448	76	41	109	154	351	119	4913	1968	372
5th B	105	386	30	54	82	219	745	346	1968	5025	273
teachers	149	1084	586	508	414	282	382	168	372	273	101

The matrix entry for row A and column B gives the total number of contacts n_{AB} measured between all individuals of classes A and B over the two days of data collection.
doi:10.1371/journal.pone.0023176.t002

Fig. 12. Number of contacts among classes (from [42])

5.3   Clustering

The Tucker decomposition produces factor matrices that represent the proximity of the elements of a dimension depending on their behavior on all the other dimensions. Therefore, classic clustering techniques can be applied on a selected factor matrix to cluster its elements.

To apply this technique, the tensor must be built with a dimension representing the samples to cluster. Enforcing the orthogonality constraint on factor matrices is of great help to separate more clearly the different clusters, so it is better to use the HOOI algorithm. The number of ranks can be chosen identically to the exploratory analysis technique.

From the result of the Tucker decomposition, it is possible to execute clustering algorithms on the factor matrix of the dimension of the elements to cluster, in order to gather similar elements. To illustrate this technique, we apply it on all the datasets presented in Sect. 5.1.

Table 2. Modeling of the tensors for the clustering experiment (the dimension on which we apply the clustering algorithm is in **bold**)

Dataset	Dimensions	Size of dimensions
Iris	Characteristics, **Samples**	4 × 150
MNIST	Pixels, Pixels, **Samples**	28 × 28 × 10 000
COIL-20 (with position)	Pixels, Pixels, Position, **Samples**	128 × 128 × 72 × 1 440
COIL-20 (without position)	Pixels, Pixels, **Samples**	128 × 128 × 1 440
Primary school	**Students**, Students, Time	242 × 242 × 208

The Table 2 summarizes the tensors built for the experiment for each dataset. The Iris dataset is embedded into a 2-order tensor, in which each sample has a vector of characteristics, namely the sepal length, the sepal width, the petal length and the petal width. For the MNIST dataset, a 3-order tensor stores the image corresponding to each sample. To reduce the size of the tensor, only 10 000 samples are kept, with 1 000 samples for each digit. For the COIL-20 dataset, we have tried two different modeling: one that includes the position of the object in a dimension, and another that uses the same representation as for the MNIST dataset, without modeling the position. Finally for the primary school dataset, we use the same modelling as for the exploratory analysis experiment.

Table 3. Result of the clustering experiment on each dataset

Dataset	Ranks used	Precision	Adjusted Rand Index
Iris	3, 3	80%	0.5623
MNIST	10, 10, 100	12.83%	0.015
COIL-20 (with position)	31, 18, 72, 20	5.63%	−0.0046
COIL-20 (without position)	31, 18, 20	42.43%	0.337
Primary school	13, 13, 4	91.38%	0.8189

From these tensors, we run the Tucker decomposition with the HOOI algorithm, and we apply the k-medoids algorithm [25] on the factor matrix corresponding to the dimension to cluster, with k being the number of classes of the dataset. The Table 3 shows the ranks used for each dataset, the precision of the clustering and the Adjusted Rand Index [21] (ARI). In this experiment, the precision is computed as follows:

$$\text{precision} = \frac{\text{number of elements correctly clustered}}{\text{total number of elements}}$$

The cluster gathering the most elements of a given class is considered as the cluster for this class. The ARI is computed as follows:

$$\text{ARI} = \frac{\sum_{i=1}^{N_c}\sum_{j=1}^{N_c}\binom{n_{i,j}}{2} - \frac{\sum_{i=1}^{N_c}\binom{n_{i,.}}{2}\sum_{j=1}^{N_c}\binom{n_{.,j}}{2}}{\binom{N}{2}}}{\frac{1}{2}\left(\sum_{i=1}^{N_c}\binom{n_{i,.}}{2} + \sum_{j=1}^{N_c}\binom{n_{.,j}}{2}\right) - \frac{\sum_{i=1}^{N_c}\binom{n_{i,.}}{2}\sum_{j=1}^{N_c}\binom{n_{.,j}}{2}}{\binom{N}{2}}}$$

with N the number of elements to cluster, N_c the number of classes and $n_{i,j}$ the value at line i and column j of the confusion matrix. $n_{.,j}$ is the sum of the column j and $n_{i,.}$ is the sum of the line i.

The clustering provides good results for the Iris (80% precision and 0.5623 ARI) and the primary school (91.38% precision and 0.8189 ARI) datasets. The confusion matrix for the Iris dataset is given in Fig. 13. As expected, the species that are not linearly separable concentrate most of the clustering errors. For the primary school dataset, we cluster only students and not the teachers, as we do not know which teacher is affected to which class.

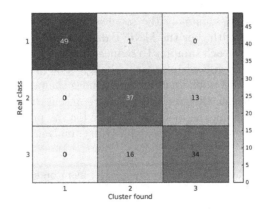

Fig. 13. Confusion matrix for the clustering of the Iris dataset

The results for the MNIST and the COIL-20 datasets are less satisfying. Indeed, for the MNIST dataset, the decomposition does not seems to be able to naturally find a pattern for each digit, and a precision of only 12.83% is obtained with an ARI of 0.015. For the COIL-20 dataset, when modeling the position into the tensor, the result is far worse (5.63% precision and -0.0046 ARI) than when the position is not represented in the tensor (42.43% precision and 0.337 ARI). Our hypothesis for this result is that the position is not a characteristics of the objects, thus the same object is never found twice with the same value on the position dimension, and the decomposition has more difficulties to find patterns in these conditions. Similarly, when the position is not represented as a dimension of the tensor, the clusters are not well defined because each sample concerns an object and a position, and the decomposition finds patterns for both at the same time.

To conclude on this technique, the experiments showed that the modeling of the tensor impacts the result, but also that the quality of the result is better if the elements of the dimension to cluster share a similar behavior on the other dimensions.

5.4   Classification

With the Tucker decomposition, it is possible to classify new elements by first building a model from elements with known class, and then by sending the new element into the same space as the model to be able to compare it with existing classes and to choose the most fitting one.

In this use case, the Tucker decomposition is used to build a model from training data. To do so, the modeling of the data into a tensor must have a dimension to represent the existing classes, that is used to indicate at which class each sub-tensor belongs. For example, for the MNIST dataset, a 4-order training tensor $\mathcal{MNIST}_{train} \in \mathbb{R}^{p1 \times p2 \times s \times c}$ can be built, with two dimensions to represent the pixels ($p1$ and $p2$), one dimension to represent the samples (s), and a last one for the digits written on images (c). The values of the tensor are the values of the corresponding pixels.

Once the training tensor is built, the Tucker decomposition can be applied to produce the model. In this use case, it is better to use an algorithm enforcing the orthogonality constraint such as the HOOI one. With our MNIST example, we obtain the following model:

$$\mathcal{MNIST}_{train} \simeq \mathcal{G} \times_1 \mathbf{P1} \times_2 \mathbf{P2} \times_3 \mathbf{S} \times_4 \mathbf{C}$$

with $\mathcal{G} \in \mathbb{R}^{R_{p1} \times R_{p2} \times R_s \times R_c}$ the core tensor, $\mathbf{P1} \in \mathbb{R}^{p1 \times R_{p1}}$ and $\mathbf{P2} \in \mathbb{R}^{p2 \times R_{p2}}$ the factor matrices for the two pixel dimensions, $\mathbf{S} \in \mathbb{R}^{s \times R_s}$ the factor matrix for the sample dimension, and $\mathbf{C} \in \mathbb{R}^{c \times R_c}$ the factor matrix for the class dimension.

The goal of a classification task is to deduce the class of a new element. To continue with the MNIST example, it consists in deducing the digit written on a new image. We consider a new image as a matrix $\mathbf{I} \in \mathbb{R}^{p1 \times p2}$. To classify this image according to the model, both must be in a comparable space. To do so, we rely on the relation among the input tensor, the core tensor and the factor matrices, that implies that the core tensor can be obtained from the input tensor by applying successive mode-n products on the input tensor and each factor matrix transposed. More formally, this relation can be summarized as follows:

$$\mathcal{G} \simeq \mathcal{MNIST}_{train} \times_4 \mathbf{C}^T \times_3 \mathbf{S}^T \times_2 \mathbf{P2}^T \times_1 \mathbf{P1}^T$$

By exploiting this relation, the matrix $\mathbf{I}$ can be partially sent into the same space as the model, by applying the mode-n product on known dimensions, namely the first and the second that represent the pixels. To be closer to the model, the matrix $\mathbf{I}$ can be considered as a 4-order tensor $\mathcal{I} \in \mathbb{R}^{p1 \times p2 \times 1 \times 1}$, by adding two dimensions of size 1:

$$\mathcal{G}_{partial} = \mathcal{I} \times_2 \mathbf{P2}^T \times_1 \mathbf{P1}^T$$

With this representation, it is not possible to fully send the element to classify in the same space as the model. Indeed, the size of the dimensions s and c does not match the size of the dimensions 3 and 4 of $\mathcal{I}$. In order to solve this problem on the third dimension, rather than only simulating a dimension of size 1, we duplicate the matrix $\mathbf{I}$ s times to obtain the 4-order tensor $\mathcal{I} \in \mathbb{R}^{p1 \times p2 \times s \times 1}$ and to be able to use one more factor matrix to send the element in the same space as the model[3]:

$$\mathcal{G}_{partial} = \mathcal{I} \times_3 \mathbf{S}^T \times_2 \mathbf{P2}^T \times_1 \mathbf{P1}^T$$

To finally classify the element, we compare $\mathcal{G}_{partial} \in \mathbb{R}^{R_{p1} \times R_{p2} \times R_s \times 1}$ with $\mathcal{G} \in \mathbb{R}^{R_{p1} \times R_{p2} \times R_s \times R_c}$ by keeping only one class at a time in $\mathcal{G}$. To do so, for each class i we use the product mode-4 on the core tensor $\mathcal{G}$ and the column vector i of factor matrix $\mathbf{C}$ to produce a class specific core tensor $\mathcal{G}_i \in \mathbb{R}^{R_{p1} \times R_{p2} \times R_s \times 1}$:

$$\mathcal{G}_i = \mathcal{G} \times_4 \mathbf{c}_i$$

As $\mathcal{G}_i$ and $\mathcal{G}_{partial}$ are now of the same size and in the same space, they can be compared with the Frobenius norm applied on the difference of the two tensors. The class for which the Frobenius norm is the lowest (i.e., for which the two tensors are the closest) can be considered as the class of the element. The classification process is summarized in Algorithm 3 for a N-order training tensor and a M-order element to classify.

---

[3] Most of the works presenting the classification technique do not perform a duplication, and directly compare the partial core tensor against each sample and each class [5,11]. It is less efficient as it implies at most $s \times c$ comparisons, while duplicating the element reduce the number of comparisons to c. During our experiments, we find it more efficient to duplicate the element, as it allows to compare a unified pattern of a class with the sample without focusing on an outlier that could negatively impact the result.

Algorithm 3. Classification process

Require: Training tensor $\mathcal{X}_{train} \in \mathbb{R}^{I_1 \times I_2 \times \cdots \times I_N}$ and the element to classify $\mathcal{E} \in \mathbb{R}^{I_1 \times I_2 \times \cdots \times I_M}$ with $M \leq (N-1)$ and I_N the number of classes
Ensure: Class c, the best matching class for $\mathcal{E}$
1: Initialize $\mathcal{G} \in \mathbb{R}^{R_1 \times R_2 \times \cdots \times R_N}$, $\mathbf{U}^{(1)}, \ldots, \mathbf{U}^{(N)} \leftarrow HOOI(\mathcal{X}_{train}, R_1, \ldots, R_N)$
2: **for** $n = M + 1, \ldots, N - 1$ **do**
3:   $\mathcal{E}(:, \ldots, 1 : I_n) \leftarrow$ repeat $\mathcal{E}$ I_n times
4: **end for**
5: $\mathcal{G}_{partial} \leftarrow \mathcal{E}$
6: **for** $n = 1, \ldots, N - 1$ **do**
7:   $\mathcal{G}_{partial} \leftarrow \mathcal{G}_{partial} \times_n \mathbf{U}^{(n)T}$
8: **end for**
9: best_result $\leftarrow$ max(Double)
10: **for** $n = 1, \ldots, I_N$ **do**
11:   $\mathcal{G}_i \leftarrow \mathcal{G} \times_N \mathbf{u}_i^{(N)}$
12:   result $\leftarrow \|\mathcal{G}_i - \mathcal{G}_{partial}\|_F$
13:   **if** result $<$ best_result **then**
14:     best_result $\leftarrow$ result
15:     $c \leftarrow n$
16:   **end if**
17: **end for**

Table 4. Modeling of the tensors for the classification experiment (the dimension that holds information about classes is in **bold**)

Dataset	Dimensions	Size of dimensions
Iris	Characteristics, Samples, **Species**	$4 \times 50 \times 3$
MNIST	Pixels, Pixels, Samples, **Digit**	$28 \times 28 \times 8\ 000 \times 10$
COIL-20	Pixels, Pixels, Positions, **Objects**	$128 \times 128 \times 72 \times 20$
Primary school	Students (s1), Students, Time, **Class (s1)**	$242 \times 242 \times 208 \times 10$

To apply the classification technique on the dataset of Sect. 5.1, we build tensors as specified in Table 4. The MNIST tensor has 8 000 samples rather than 7 000 because the digits are not evenly distributed and some digits are represented in more than 7 000 images. The classes of the primary school dataset concern the students of the first dimension. To experiment the technique, we use the cross validation method and perform the training and the classification task 5 times. The data used for the training step are modified at each iteration to avoid overfitting bias.

The results obtained are summarized in Table 5, and the detailed metrics for each class are given in Table 6. Most of the samples are correctly classified for all the datasets. For the MNIST dataset, there is a great improvement compared to the clustering technique: the global precision is almost 8 times better. It indicates

Table 5. Result of the classification experiment on each dataset. For the COIL-20 dataset, "without position" indicates that images were classified only regarding objects . and "with position" indicates that the images were classified regarding positions and objects

Dataset	Training samples per class	Ranks used	Precision
Iris	20	2, 3, 3	88.22%
MNIST	2 000	9, 8, 1, 10	81.02%
COIL-20 (with position)	20	20, 20, 72, 20	100%
COIL-20 (without position)	40	20, 20, 72, 20	61.75%
Primary school	10	10, 10, 4, 10	94.81%

Table 6. Detailed metrics for each class of each dataset for the classification technique

Class	Precision	Recall	F1-score	Class	Precision	Recall	F1-score
Iris dataset				COIL-20 dataset (with position)			
Setosa	94.29%	100%	96.67%	1	100%	100%	100%
Versicolor	79.53%	92.67%	85.28%	2	100%	100%	100%
Virginica	95.56%	72%	82.08%	3	100%	100%	100%
MNIST dataset				4	100%	100%	100%
0	91.4%	86.83%	89.05%	5	100%	100%	100%
1	74.41%	96.28%	83.92%	6	100%	100%	100%
2	87.07%	75.68%	80.97%	7	100%	100%	100%
3	76.1%	77.62%	76.83%	8	100%	100%	100%
4	79.87%	81.18%	80.51%	9	100%	100%	100%
5	72.73%	67.65%	70.06%	10	100%	100%	100%
6	87.84%	87.08%	87.45%	11	100%	100%	100%
7	90.43%	84.08%	87.13%	12	100%	100%	100%
8	79.62%	72.33%	75.8%	13	100%	100%	100%
9	74.62%	77.85%	76.2%	14	100%	100%	100%
Primary school dataset				15	100%	100%	100%
1A	100%	95.38%	97.53%	16	100%	100%	100%
1B	100%	100%	100%	17	100%	100%	100%
2A	100%	88.77%	94.01%	18	100%	100%	100%
2B	100%	93.33%	96.42%	19	100%	100%	100%
3A	100%	96.92%	98.84%	20	100%	100%	100%
3B	100%	90%	94.62%				
4A	100%	100%	100%				
4B	68.15%	100%	80.31%				
5A	100%	93.33%	96.44%				
5B	100%	90.65%	94.98%				

that, when integrating more contextual information into the tensor (in this case, the digit written), the Tucker decomposition can find patterns more easily.

With the COIL-20 dataset, it is possible to illustrate a useful mechanism of the classification performed from a model obtained with the Tucker decomposition. Indeed, it can be used to classify an element according to several different class dimensions rather than just one. In the COIL-20 dataset, each image represents an object, but also a specific position. To illustrate this behavior, we classify the test images according to the object that they represent but also to their position. The Fig. 14 shows the confusion matrices obtained for this experiment. The objects are better recognized, and the position is found for 94.81% of the test images with a precision of ± 5°.

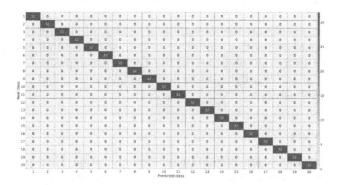

(a) Confusion matrix for the objects

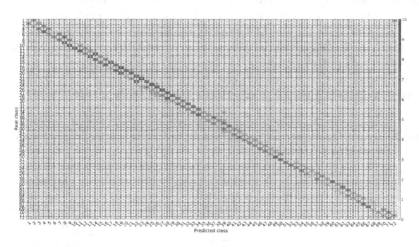

(b) Confusion matrix for the positions

Fig. 14. The confusion matrices obtained when classifying elements of the COIL-20 dataset according to the object that they represent and their position

The Tucker decomposition shows promising results when using it in classification tasks. It can be used to classify a new element depending on one or several parameters, only by using the model produced by the decomposition algorithm.

6 Robustness of the Tucker Decomposition

It is important to evaluate the robustness of an algorithm, as it gives information about the perturbations that can occur in data without significantly impacting the result. We study the robustness of the Tucker decomposition when it is used for clustering or for classifying tasks with missing values, and show that it has a fairly good robustness.

6.1 Clustering

For testing the robustness of the Tucker decomposition when performing clustering tasks with missing data, the primary school dataset is used as it presents the best results for the clustering task with all the data. 5 students are selected from each class, and 10% of data are randomly removed at each iteration only for those students, for 9 iterations. Thus, the clustering is performed with 90% of the students' data for the first execution and with 10% of the data for the last execution. Table 7 gives the precision and the ARI for the whole data and for the selected students, and Fig. 15 shows the confusion matrices of the result of the experiment. Confusion matrices on the left are for the whole dataset and confusion matrices on the right are for the selected students only.

Table 7. Result of the clustering experiment on the primary school dataset with missing data for selected students

% of missing data	Global precision	Global Adjusted Rand Index	Precision for selected students	Adjusted Rand Index for selected students
10%	84.48%	0.7782	86%	0.7915
20%	80.17%	0.6781	82%	0.6247
30%	88.79%	0.7577	84%	0.5947
40%	89.66%	0.7806	84%	0.5947
50%	86.21%	0.6641	62%	0.1534
60%	86.21%	0.6502	48%	0.0891
70%	72.84%	0.3701	10%	0
80%	65.52%	0.3335	10%	0
90%	73.28%	0.3638	10%	0

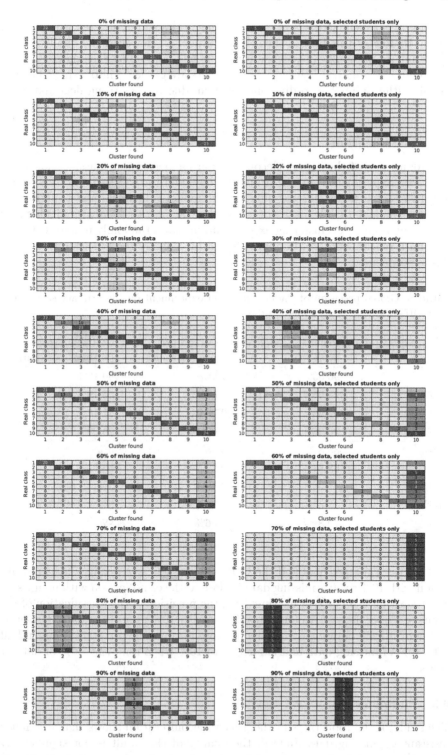

Fig. 15. Confusion matrices for the clustering of the primary school dataset with missing data for selected students

This experiment shows that the clustering is almost not affected when removing up to 40% of the data of the selected students. However, the quality of the clustering drops significantly when removing 50% and 60% of the data. For 70% or more missing data, all the selected students are gathered into a single cluster: the algorithm is unable to differentiate them due to the strong degradation of the data.

6.2   Classification

The robustness of the Tucker decomposition for performing classification tasks have been experimented on the primary school and the COIL-20 datasets. Compared to Sect. 5.4, the step for building the model is identical and is performed with the whole data. The data are removed in the test set evenly for all elements to classify. As for the clustering experiment, data are randomly removed 10% by 10% until reaching 90% of missing data. Table 8 shows the precision results for this experiment.

Table 8. Global precision of the classification experiment on the primary school and the COIL-20 datasets with missing data

% of missing data	Primary school precision	COIL-20 precision (with position)	COIL-20 precision (without position)
10%	96.63%	100%	59.12%
20%	95.98%	99.81%	56.19%
30%	94.63%	97.84%	41.16%
40%	90.96%	48.06%	25.09%
50%	83.87%	9.78%	5%
60%	68.81%	0.34%	5%
70%	45.94%	4.18%	5%
80%	23.87%	7.22%	5%
90%	12.89%	7.06%	5%

For the primary school and the COIL-20 datasets, the quality of the classification with 10% to 30% missing data is close to the classification with no missing data. Starting from 50% of missing data, the precision of the classification of the primary school dataset declines steadily at each 10% removal of data. However, for the COIL-20 dataset, the decline is more abrupt: the precision is cut in half with 40% of missing data and is inferior to 10% when removing 50% of data or more.

6.3   Summary

The study of the robustness of the Tucker decomposition shows that it is fairly resistant to missing data. Indeed, the quality of the result is not significantly reduced when removing up to 30% of data. Furthermore, when comparing results obtained from the primary school dataset and from the COIL-20 dataset, the lowering of quality is less abrupt with the primary school dataset, thus indicating that it is a well suited data mining technique when working with sparse data that present imperfections of the real world (e.g., students of a same class do not act identically).

7 Conclusion

To conclude, the Tucker decomposition is a useful data mining algorithm, robust to missing data. Indeed, it can be used to perform exploratory analysis on data, in order to retrieve patterns that give insights regarding elements of a given dimension, and regarding relationship of elements among dimensions. It can also be used to cluster elements of a dimension when they behave similarly on all the other dimensions, or to produce a model allowing to classify new data according to one or several characteristics.

Both the HOOI and the HALS-NTD algorithms are useful for these techniques, as the non-negativity constraint of the HALS-NTD greatly helps when interpreting results of exploratory analysis, while the orthogonality constraint of the HOOI algorithm is efficient to cluster or classify data. However, the HALS-NTD algorithm is less known than the HOOI one, and in consequence it has almost never been implemented. We plan to integrate these Tucker algorithms to the Tensor Data Model, and to optimize them in order to allow their execution on large tensors, as we did for the CANDECOMP/PARAFAC decomposition [13]. Indeed, real data can create such tensors, that emphasis the need for optimized algorithms regarding the space and the execution time.

We also plan to improve data mining techniques based on the experiments made on this article, for example to consider a proximity among elements of a dimension (e.g., two consecutive time slices on a temporal dimension are closer than non-consecutive time slices), or to perform a coupled decomposition, i.e., a decomposition with two tensors that share at least one dimension.

References

1. Al-Sharoa, E., Al-Khassaweneh, M., Aviyente, S.: A tensor based framework for community detection in dynamic networks. In: 2017 IEEE International Conference on Acoustics, Speech and Signal Processing (ICASSP), pp. 2312–2316. IEEE (2017)
2. Angles, R., Arenas, M., Barceló, P., Hogan, A., Reutter, J., Vrgoč, D.: Foundations of modern query languages for graph databases. ACM Comput. Surv. (CSUR) **50**(5), 1–40 (2017)
3. Araujo, M., et al.: Com2: fast automatic discovery of temporal ('Comet') communities. In: Tseng, V.S., Ho, T.B., Zhou, Z.-H., Chen, A.L.P., Kao, H.-Y. (eds.) PAKDD 2014. LNCS (LNAI), vol. 8444, pp. 271–283. Springer, Cham (2014). https://doi.org/10.1007/978-3-319-06605-9_23

4. Atikoglu, B., Xu, Y., Frachtenberg, E., Jiang, S., Paleczny, M.: Workload analysis of a large-scale key-value store. In: ACM SIGMETRICS Performance Evaluation Review, vol. 40, pp. 53–64. ACM (2012)
5. Brandoni, D., Simoncini, V.: Tensor-train decomposition for image recognition. Calcolo **57**, 1–24 (2020)
6. Chachlakis, D.G., Prater-Bennette, A., Markopoulos, P.P.: L1-norm tucker tensor decomposition. IEEE Access **7**, 178454–178465 (2019)
7. Cichocki, A., Zdunek, R., Phan, A.H., Amari, S.: Nonnegative Matrix and Tensor Factorizations: Applications to Exploratory Multi-way Data Analysis and Blind Source Separation. Wiley, Chichester (2009)
8. De Lathauwer, L., De Moor, B., Vandewalle, J.: A multilinear singular value decomposition. SIAM J. Matrix Anal. Appl. **21**(4), 1253–1278 (2000)
9. Deng, L.: The MNIST database of handwritten digit images for machine learning research [best of the web]. IEEE Signal Process. Mag. **29**(6), 141–142 (2012)
10. Duan, L., Xiao, C., Li, M., Ding, M., Yang, C.: a-tucker: fast input-adaptive and matricization-free tucker decomposition of higher-order tensors on GPUs. CCF Trans. High Perform. Comput. **5**(1), 12–25 (2023)
11. Eldén, L.: Matrix methods in data mining and pattern recognition. In: SIAM (2007)
12. Fernandes, S., Fanaee-T, H., Gama, J.: Tensor decomposition for analysing time-evolving social networks: an overview. Artif. Intell. Rev. **54**, 2891–2916 (2021)
13. Gillet, A., Leclercq, É., Cullot, N.: MuLOT: multi-level optimization of the canonical polyadic tensor decomposition at large-scale. In: Bellatreche, L., Dumas, M., Karras, P., Matulevičius, R. (eds.) ADBIS 2021. LNCS, vol. 12843, pp. 198–212. Springer, Cham (2021). https://doi.org/10.1007/978-3-030-82472-3_15
14. Gillet, A., Leclercq, E., Sautot, L.: The tucker tensor decomposition for data analysis: capabilities and advantages. In: 38ème Conférence sur la Gestion de Données (BDA) (2022)
15. Gillet, A., Leclercq, É., Savonnet, M., Cullot, N.: Empowering big data analytics with polystore and strongly typed functional queries. In: Symposium on International Database Engineering & Applications, pp. 1–10 (2020)
16. Gray, J., et al.: Data cube: a relational aggregation operator generalizing group-by, cross-tab, and sub-totals. Data Min. Knowl. Disc. **1**(1), 29–53 (1997)
17. Hamilton, J.D.: Time Series Analysis. Princeton University Press, Princeton (2020)
18. Hore, V., et al.: Tensor decomposition for multiple-tissue gene expression experiments. Nat. Genet. **48**(9), 1094–1100 (2016)
19. Hou, Z., Li, W., Tao, R., Du, Q.: Three-order tucker decomposition and reconstruction detector for unsupervised hyperspectral change detection. IEEE J. Sel. Top. Appl. Earth Obs. Remote Sens. **14**, 6194–6205 (2021)
20. Huang, H., Ding, C., Luo, D., Li, T.: Simultaneous tensor subspace selection and clustering: the equivalence of high order SVD and k-means clustering. In: Proceedings of the 14th ACM SIGKDD International Conference on Knowledge Discovery and Data Mining, pp. 327–335 (2008)
21. Hubert, L., Arabie, P.: Comparing partitions. J. Classif. **2**, 193–218 (1985)
22. Jang, J.G., Kang, U.: D-tucker: fast and memory-efficient tucker decomposition for dense tensors. In: 2020 IEEE 36th International Conference on Data Engineering (ICDE), pp. 1850–1853. IEEE (2020)
23. Jang, J.G., Kang, U.: Static and streaming tucker decomposition for dense tensors. ACM Trans. Knowl. Discov. Data **17**(5), 1–34 (2023)
24. Kanellakis, P.C.: Elements of relational database theory. In: Formal Models and Semantics, pp. 1073–1156. Elsevier (1990)

25. Kaufman, L., Rousseeuw, P.J.: Finding Groups in Data: An Introduction to Cluster Analysis. Wiley, Hoboken (2009)
26. Kivelä, M., Arenas, A., Barthelemy, M., Gleeson, J.P., Moreno, Y., Porter, M.A.: Multilayer networks. J. Complex Netw. **2**(3), 203–271 (2014)
27. Kolda, T.G., Bader, B.W.: Tensor decompositions and applications. SIAM Rev. **51**(3), 455–500 (2009)
28. Leclercq, É., Gillet, A., Grison, T., Savonnet, M.: Polystore and tensor data model for logical data independence and impedance mismatch in big data analytics. In: Hameurlain, A., Wagner, R. (eds.) Transactions on Large-Scale Data- and Knowledge-Centered Systems XLII. LNCS, vol. 11860, pp. 51–90. Springer, Heidelberg (2019). https://doi.org/10.1007/978-3-662-60531-8_3
29. Lee, J., Chon, K.W., Kim, M.S.: A GPU-based tensor decomposition method for large-scale tensors. In: 2023 IEEE International Conference on Big Data and Smart Computing (BigComp), pp. 77–80. IEEE (2023)
30. Li, L., Lin, X., Liu, H., Lu, W., Zhou, B., Zhu, J.: Displacement data imputation in urban internet of things system based on tucker decomposition with l2 regularization. IEEE Internet Things J. **9**(15), 13315–13326 (2022)
31. Nene, S.A., Nayar, S.K., Murase, H., et al.: Columbia object image library (coil-20) (1996)
32. Osman, A.S.: Data mining techniques. Int. J. Data Sci. Res. **2** (2019)
33. Pandey, S.K., Shekhawat, H.S., Prasanna, S.: Attention gated tensor neural network architectures for speech emotion recognition. Biomed. Signal Process. Control **71**, 103173 (2022)
34. Papalexakis, E.E., Akoglu, L., Ience, D.: Do more views of a graph help? Community detection and clustering in multi-graphs. In: International Conference on Information Fusion, pp. 899–905. IEEE (2013)
35. Papalexakis, E.E., Faloutsos, C., Sidiropoulos, N.D.: Tensors for data mining and data fusion: models, applications, and scalable algorithms. Trans. Intell. Syst. Technol. (TIST) **8**(2), 16 (2016)
36. Petersohn, D., et al.: Towards scalable dataframe systems. arXiv preprint arXiv:2001.00888 (2020)
37. Phan, A.H., Cichocki, A.: Extended HALS algorithm for nonnegative tucker decomposition and its applications for multiway analysis and classification. Neurocomputing **74**(11), 1956–1969 (2011)
38. Romeo, S., Tagarelli, A., Ienco, D.: Semantic-based multilingual document clustering via tensor modeling. In: EMNLP: Empirical Methods in Natural Language Processing, pp. 600–609 (2014)
39. Rush, A.: Tensor Considered Harmful. Technical report, Harvard NLP (2010). http://nlp.seas.harvard.edu/NamedTensor
40. Shao, P., Zhang, D., Yang, G., Tao, J., Che, F., Liu, T.: Tucker decomposition-based temporal knowledge graph completion. Knowl.-Based Syst. **238**, 107841 (2022)
41. Sidiropoulos, N.D., De Lathauwer, L., Fu, X., Huang, K., Papalexakis, E.E., Faloutsos, C.: Tensor decomposition for signal processing and machine learning. Trans. Signal Process **65**(13), 3551–3582 (2017)
42. Stehlé, J., et al.: High-resolution measurements of face-to-face contact patterns in a primary school. PLoS ONE **6**(8), e23176 (2011)
43. Sun, J., Papadimitriou, S., Lin, C.Y., Cao, N., Liu, S., Qian, W.: Multivis: content-based social network exploration through multi-way visual analysis. In: Proceedings of the 2009 SIAM International Conference on Data Mining, pp. 1064–1075. SIAM (2009)

44. Sun, J., Tao, D., Faloutsos, C.: Beyond streams and graphs: dynamic tensor analysis. In: ACM SIGKDD International Conference on Knowledge Discovery and Data Mining, pp. 374–383. ACM (2006)
45. Tucker, L.R.: Some mathematical notes on three-mode factor analysis. Psychometrika **31**(3), 279–311 (1966)
46. Yang, K., et al.: Tagited: predictive task guided tensor decomposition for representation learning from electronic health records. In: Proceedings of the Thirty-First AAAI Conference on Artificial Intelligence (2017)
47. Zhou, G., Cichocki, A., Zhao, Q., Xie, S.: Efficient nonnegative tucker decompositions: algorithms and uniqueness. IEEE Trans. Image Process. **24**(12), 4990–5003 (2015)

Challenges for Healthcare Data Analytics Over Knowledge Graphs

Maria-Esther Vidal[1,2,3]([ID]), Emetis Niazmand[1,2] [ID], Philipp D. Rohde[1,2] [ID], Enrique Iglesias[1,3] [ID], and Ahmad Sakor[3] [ID]

[1] Leibniz University, Hannover, Germany
[2] TIB Leibniz Information Centre for Science and Technology, Hannover, Germany
`{Maria-Esther.Vidal,Emetis.Niazmand,Philipp.Rohde}@tib.eu`
[3] L3S Research Center, Hannover, Germany
`iglesias@l3s.de`

Abstract. Over the past decade, the volume of data has experienced a significant increase, and this growth is projected to accelerate in the coming years. Within the healthcare sector, various methods (such as liquid biopsies, medical images, and genome sequencing) generate substantial amounts of data, which can lead to the discovery of new biomarkers. Analyzing big data in healthcare holds the potential to advance precise diagnostics and effective treatments. However, healthcare data faces several complexity challenges, including volume, variety, and veracity, which necessitate innovative techniques for data management and knowledge discovery to ensure accurate insights and informed decision-making. This paper summarizes the results presented in the invited talk at BDA 2022 and addresses these challenges by proposing a knowledge-driven framework able to handle complexity issues associated with big data and their impact on analytics. In particular, we propose the use of Knowledge Graphs (KGs) as data structures that enable the integration of diverse healthcare data and facilitate the merging of data with ontologies that describe their meaning. We show the benefits of leveraging KGs to uncover patterns and associations among entities. Specifically, we illustrate the application of rule mining tasks that enhance the understanding of the role of biomarkers and previous cancers in lung cancer.

Keywords: Healthcare Data Analytics · Knowledge Graphs · Semantic Data Integration

1 Introduction

The healthcare sector currently faces challenges due to data silos, where large amounts of heterogeneous data are stored in fragmented structured or semi-structured formats. This fragmentation hinders the combination, analysis, and re-use of data, preventing the generation of valuable insights for decision-making in healthcare. The data silos exist for various reasons, including the volume and variety of data, restrictive data access schemes imposed by providers, data

A. Hameurlain et al. (Eds.): *TLDKS LIV*, LNCS 14160, pp. 89–118, 2023.
https://doi.org/10.1007/978-3-662-68014-8_4

sovereignty, privacy and security concerns, as well as trust, legal, ethical compliance (e.g., GDPR), missing data harmonization, low interoperability, and technical limitations. All of these factors pose significant challenges to data integration and analysis, limiting the discovery of breakthrough knowledge and the development of new technologies. Consequently, data silos restrict healthcare providers' ability to make data-driven decisions, thereby impacting civil society.

Knowledge graphs (KGs) have emerged as effective data structures for representing the convergence of data and knowledge from diverse sources [34]. KGs can be defined as data integration systems (DISs) [45], comprising a unified schema, data sources, and mapping rules that define the concepts within the schema and link them to the data sources. Declarative definitions of KGs promote modularity and reusability, allowing users to trace the KG creation process, thereby enhancing transparency and maintenance. KGs provide expressive data structures to model integrated data and metadata, and their declarative specifications can be explored, validated, and traced using existing queries, e.g., SPARQL or SQL.

The scientific community has dedicated considerable attention to the problem of data integration [32,33] and to the development of frameworks for data integration [12,13,75], data ecosystems [27], as well as to the formalization of data integration systems [45,52], and the definition of standards to define mapping rules (e.g., R2RML [14] and RML [17]). Existing formalisms allow for declaratively defining pipelines to create KGs in real-world applications (e.g., biomedical area [3,7,41,62,66,70] and energy [38]); they have allowed for establishing transparent, maintainable, and traceable processes for KG creation, as well as for providing the basis for analytical frameworks developed on top of KGs [10,59,62].

Research Goal: This paper addresses the problem of semantic data integration in healthcare and presents a knowledge-driven pipeline capable of merging data into a federation of KGs. Data integration involves the task of determining whether two entities from a collection of data sources (both structured and unstructured) refer to the same real-world entity or not, and their alignment. This process requires the identification and resolution of interoperability conflicts that arise when integrating different data sets. By recognizing and resolving these conflicts, data integration enables the harmonization of disparate data sources, facilitating a unified view of the underlying real-world entities.

In the healthcare domain, decisions are made based on evidences and the clinical experts' experience. Thus, following a classical research methodology, a hypothesis, once generated, must be tested, validated, or refuted. Data-driven analysis provides the basis for providing evidences and novel questions, empowering thus, the power of traditional experimental methods [35]. However, data-driven approaches resort to healthcare data which is usually scattered across heterogeneous data sources and whose meaning is fragmented in heterogeneous vocabularies. KGs have gained attention as expressive data structures that enable data integration, as well as the harmonization of fragmented knowledge [11].

Proposed Solution - A Knowledge Graph-based Approach: Integrating data in a Knowledge Graph (KG) facilitates the provision of comprehensive

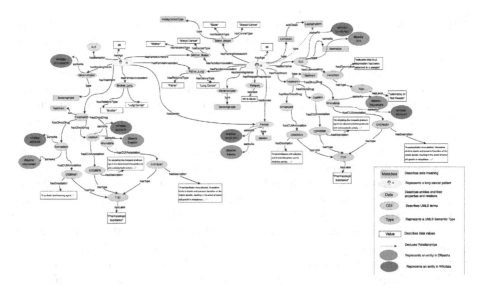

Fig. 1. A portion of a Knowledge Graph for Lung Cancer (KG4C)

descriptions for integrated entities and promotes a shared understanding of their meaning. Additionally, by leveraging KGs, inductive learning methods can be developed to unveil patterns that elucidate associations among entities. In this paper, we explain a knowledge-driven pipeline where data is collected from heterogeneous data sources, and an integrated schema is utilized to populate a KG. This pipeline has been followed by Aisopos et al. [3], Sakor et al. [62], Torrente et al. [68], and Vidal et al. [70] to provide unified views of heterogeneous biomedical data sources and pave the way for the development of discovery methods for data analysis. To illustrate the power of KGs, we employ the symbolic learning methods proposed by Lajus et al. [42] and show how the discovered rules shed light on the significance of cancer history and biomarkers in lung cancer patients. We discuss the expressive power of the extracted patterns, as well as the validity of the results with respect to outcomes reported in the literature. Finally, we close the paper with an outlook to the future discussing grand challenges required to be addressed to provide frameworks for analytical methods.

2   Preliminaries

Knowledge Graph: A Knowledge Graph $\mathcal{G}$ is defined over a set Con of countable infinite constants; it corresponds to a directed edge-labeled graph tuple $\mathcal{G} = \langle V, E, L \rangle$, where: **a)** $V \subseteq Con$ is a set of nodes; **b)** $L \subseteq Con$ is a set of edge labels; and **c)** $E \subseteq V \times L \times V$ is a set of edges. Figure 1 shows a portion of a KG for Cancer (KG4C), where nodes represent cancer patients and edges their properties. The figure depicts nodes in different colors representing metadata, cancer patients, values of properties and attributes, CUIs or terms from

UMLS[1], semantic types, data values, deduced relations, and entities from DBpedia and Wikidata. KGs represent factual statements using a graph data model, and metadata and data can be empowered with inference to deduce new facts. Inferred links are depicted in Fig. 1 with dashed red arrows. These new statements are deduced based on the entailment regimes of ontological formalisms utilized to describe the ontology of the KG. In case of KG4C, the entailment regimes are defined on the axioms that define the semantics of RDFS; they define subclass, subproperties, domain and range, and typing [34]. Additionally, the property sameAs, from the Web Ontology Language (OWL), is utilized to represent the logical equivalence between two entities. The entailment regime of the property sameAs is defined based on the Leibniz Inference Rule [29]; it enables to deduce that two entities related with the property sameAs have the same properties. For example, Fig. 2 presents a portion of KG4C where the entities `Vinorelbine`, `dbpedia:Vinorelbine`, and `wikidata:Q420532` represent the chemotherapy drug Vinorelbine; they are related via the sameAs predicate. As depicted, applying the entailment regime of sameAs, it is deduced that the three entities participate– in the subject and object position– in the same triples.

Knowledge Graph Specification: A KG $\mathcal{G}$ can be defined as a data integration system $DIS_{\mathcal{G}} = \langle O, S, M \rangle$. Here, O represents a unified ontology consisting of classes and properties, S denotes a set of data sources, and M corresponds to mapping rules or assertions formulated as conjunctive queries over the sources in S. By executing these mapping rules M over the data sources in S, instances are generated in $\mathcal{G}$. The rules in M (Namici et al. [52]) can be represented as Horn clauses, such as $body(\overline{X}) : -head(\overline{Y})$, following the Global As View (GAV) approach [45]. The entities that correspond to instances of classes in O are the nodes in V, while the edges in E represent properties of these entities.

Knowledge Graph Creation: A pipeline for creating a KG corresponds to a partial order of the mapping rules in M that represents an execution plan for the mappings. Iglesias et al. [37] propose an approach to generate logical bushy-plans of the mapping rules which can be translated into bash commands of an operating system and allow the efficient execution of a KG creation pipeline.

Federation of Knowledge Graphs: A federation is a set of KGs that shared common entities but probably represent different perspectives of the shared entities. For example, Fig. 2 depicts a portion of the KGs that compose a federation for Cancer. Each KGs is autonomous and accessible via a SPARQL endpoint. A federated query engine is a system that enables the execution of queries over a group of KGs that are federated and distributed. These engines are usually empowered with query processing methods to decompose an input query into sub-queries executable by at least one of the KGs in the federation and select the KGs more suitable for executing each sub-query. They are also responsible for finding and managing efficient plans that minimize the cost of merging data collected from the selected KGs.

---

[1] Unified Medical Language System https://www.nlm.nih.gov/research/umls/index.html.

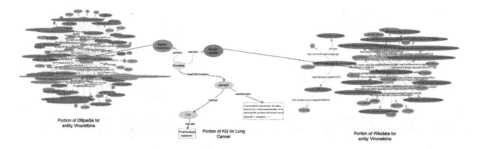

Fig. 2. A portion of the Federation of the Knowledge Graphs for Cancer. The federation is composed of DBpedia [44], Wikidata [71], and the Cancer KG [3].

3   A Pipeline for Data Integration and Knowledge Graphs

The integration of diverse data sources into a federation of Knowledge Graphs (KGs) is made possible through the declarative definition of data integration systems, which specify the KGs within the federation. In Fig. 3, we present a pipeline that incorporates multiple computational frameworks, enabling the resolution of interoperability issues found in the heterogeneous data sources. These issues are addressed by incorporating factual statements from the data sources into the KGs of the federation. After the data is ingested, Named Entity Recognition (NER) and Entity Linking (EL) tasks are performed to identify alignments across different entities that correspond to the same real-world entity. These alignments establish links between the entity "Vinorelbine" and their corresponding entities in DBpedia, Wikidata, and UMLS. Furthermore, mapping rules are defined to specify the correspondences between concepts (e.g., classes and properties) of a KG ontology and the data sources. In the pipeline described in this paper, the mapping rules are expressed using the RDF Mapping Language (RML) [15]. The KG creation engine executes these mapping rules to generate the KG. The links between the KGs are represented using the "sameAs" property. To ensure KG correctness, integrity constraints are expressed as shapes using the SHACL language[2]. Shapes allow for the representation of constraints over the properties of a class or between properties that connect two classes [34]. A federated query engine is responsible for executing queries by exploring the KGs in the federation. Lastly, symbolic learning mines patterns in the form of Horn clauses that model meaningful properties within a KG [25].

Data Ingestion: Data sources are ingested in various formats (e.g., tabular, semi-structured, or unstructured sources); they can also be modeled using different data models (e.g., relational, graph, or hierarchical data models). Data sources are stored in a research data management repository and their metadata described using controlled vocabularies (e.g., DCAT[3]). To respect data access regulations and privacy policies, the clinical data sources are only accessible by

---

[2] https://www.w3.org/TR/shacl/.
[3] https://www.w3.org/TR/vocab-dcat-3/.

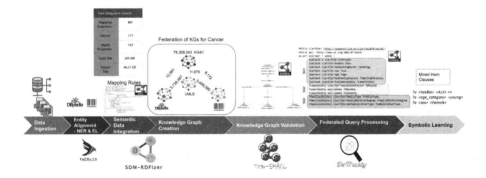

Fig. 3. Pipeline to Create a Federation of KGs for Cancer.

authorized users, while open data sources are publicly accessible. An instance of the Leibniz Data Manager (LDM) [9] is the repository of KG4C data sources.

Healthcare data is heterogeneous and usually scattered across many different sources. Moreover, biomedical knowledge is fragmented in various vocabularies [11]. This pipeline is domain- and application-agnostic; however, for the creation of KG4C, the following data sources are considered: **i)** Clinical records: are presented in one universal table comprising more than 1,500 attributes describing cancer patients [3] shared in the context of the EU H2020 funded project CLARIFY[4]; some attributes include short textual notes that encode meaningful information about a patient. **ii)** Publications: data sources with data referring to scientific publications. In KG4C, publications are ingested from SemMedDB [40], which includes scholarly metadata of PubMed publications, as well as fine-grained description of their abstracts based on terms and relationships from UMLS. **iii)** Drugs: data sources including drugs, side effects, toxicities, and interactions between drugs. In KG4C, NER and EL methods are employed to extract from DrugBank[5] drug-drug interactions and provide a fine-grained description of the pair of drugs that interact, and the effect and impact of the interaction. Similarly, interactions between drugs and foods are extracted, as well as absorption routes and mechanisms of action of each drugs. **iv)** Familial Cancer History: this is a tabular data source that includes, for each patient, the history of previous cancers and cancers suffered by the patient's relatives. In total, 8,493,076 records are collected from all these data sources.

Entity Alignment - Based on Named Entity Recognition (NER) and Entity Linking (EL): NER corresponds to the task of recognizing entities within a short text T, while LE links the recognized entities to equivalent entities in other knowledge bases. FALCON [63,64] implements the tasks of NER and EL, and aligns entities recognized in the attributes of the previously described data sources, to terms in UMLS, DBpedia, and Wikidata. FALCON resorts to background knowledge built on top of the UMLS, DBpedia, and Wikidata. This

---

[4] https://www.clarify2020.eu/.

[5] https://go.drugbank.com/.

Table 1. Named Entity Recognition and Linking (Accuracy). Comparison of FALCON with existing approaches. FALCON symbolic system is competitive in the tasks of named entity recognition and linking to UMLS.

	MedMentions [48]	BC5CDR [46]
SciSpacy [53]	38.8	53.9
N-GRAM TF-IDF [57]	50.9	**86.9**
FALCON [63,64]	**65.3**	80.4

background knowledge comprises resources in these KGs together with their definitions and types; it is composed of more than 50M entities. FALCON also relies on a rule-based system used to extract mentions of entities in a text and perform EL. The rule-based system is guided by a catalog of linguistic and domain-specific rules. Linguistic rules state the criteria to recognize entities in a sentence of a particular language. Domain-specific rules define what is an entity in a particular domain. FALCON rules are based on the assumption that entities have labels, definitions, and semantic types. Since the used KGs (e.g., UMLS, DBpedia, and Wikidata) can be community-maintained, the same resource may have several values of the same property (e.g., various labels or definitions). Additionally, a resource may have equivalent entities in other KGs, and inference processes enable the computation of the transitive closure over the logical equivalence that exists among equivalent entities.

FALCON is evaluated on MedMentions [48] and BC5CDR [46]. MedMentions is a corpus of 4,392 scientific papers (Titles and Abstracts) from PubMed, annotated with mentions of UMLS entities. BC5CDR comprises the abstracts of 1,500 PubMed articles annotated with the MeSH[6] vocabulary. We utilized alignments between the UMLS and MeSH vocabularies to annotate the BC5CDR dataset with entities from UMLS. FALCON is compared with **a) TF-idf**: is the widely used candidate retrieval model [4,57]. **b) SciSpacy**: is based on the Spacy library for biomedical text progressing [53]. The results are reported in Table 1. As observed, FALCON outperforms SciSpacy in the two benchmarks, while it is competitive with N-GRAM TF-IDF. The presented results support the choice of FALCON for performing NER and EL, and providing the basis for entity alignment based on UMLS annotations. Moreover, FALCON aligns entities recognized in the data sources with their corresponding entities in DBpedia and Wikidata. Figure 3 reports on the number of links between the KGs of the federation. In the KG4C federation from KG4C, there are 12,961; 8,172; and 11,679 links to DBpedia, Wikidata, and UMLS respectively. FALCON was also used to align UMLS with DBpedia and Wikidata. As a result, there are 3,739,487 links from UMLS to DBpedia, and 3,499,580 to Wikidata.

Semantic Data Integration: Entity alignments identified during NER and EL provide the basis for the semantic data integration of the data sources. Additionally, the correspondences between the data sources and the KG ontology are specified in terms of RML mapping rules. In particular, KG4C is defined in

---

[6] https://www.ncbi.nlm.nih.gov/mesh/.

terms of 1,749 mapping rules which have been designed by two knowledge engineers and reviewed by one senior researcher. These mapping rules solve various heterogeneity conflicts [70] (e.g., structuredness, schematic, domain, and representation); they are accessible via a SPARQL endpoint[7]. Thus, the definition of classes and properties can be retrieved by simply executing SPARQL queries.

Knowledge Graph Creation: A pipeline for data integration and KG creation is followed to transform the data sources in a data integration system into a KG and its links to other KGs that composed the created federation. In the case of KG4C, an ontology composed of 177 classes, 143 object properties, and 64 data type properties, provide a unified view of the concepts represented in the data sources to be integrated. Additionally, the 1,749 RML mapping rules specify the correspondences between the 179 data sources and these concepts. This pipeline is executed using the SDM-RDFizer [36] RML engine. Additionally, the mapping rule planner proposed by Iglesias et al. [37] identifies an execution plan of these 1,749 mapping rules.

An empirical evaluation was set up to understand the impact of the planner in the KG creation process; a more exhaustive empirical evaluation is reported by Iglesias et al. [37]. The experiment consists of two testbeds: **TB1)** The set of mapping rules are included in one group. **TB2)** The set of mapping rules are partitioned into eight groups; each group includes the mapping rules that define a set of related concepts, e.g., annotations, drugs, clinical records, publications, genomics data, and wearable patient profiles. These groups were created by the knowledge engineers that defined the mapping rules.

The planner identified an execution plan– in the form of busy tree– for the mappings in **TB1** and eight execution plans for the mapping rules in **TB2**. These plans were executed in an Intel(R) Xeon(R) equipped with a CPU E5-2603 v3 @ 1.60GHz 20 cores, 64GB memory and with the O.S. Ubuntu 16.04LTS. *Execution time* was measured as the elapsed time spent to evaluate each bushy tree and store the result in secondary memory. The execution time is measured using the Python library `time`. The experiments were executed five times and the average is reported; a timeout of five hours was set up. In addition to SDM-RDFizer, Morph-KGC [6] and RMLMapper [16] are included in the study.

Table 2 reports on the results of execution time (in secs.). The process of KG creation benefits significantly from executing the mapping rules in groups. This approach has been observed to reduce execution time in all the cases. Notably, the most substantial time savings are achieved when the planning process is applied to the groups already created by the experts, resulting in a reduction of up to 68.56%. These results underscore the importance of combining experts' knowledge with optimization techniques to devise plans aimed at minimizing the cost of executing complex processes.

Knowledge Graph Validation: Constraints that define the properties of classes can be categorized as intra-class and inter-class constraints. Intra-class constraints encompass aspects such as data types, cardinalities, functional

---

[7] https://labs.tib.eu/sdm/clarify_mappings_and_ontology/sparql.

Table 2. Performance of the KG Creation and the effect of planning mapping rules. A comparison was conducted to evaluate the effects of planning the execution of mapping rules during the process of Knowledge Graph (KG) creation. Two testbeds were used: **TB1**, which included all the mapping rules in a single group, and **TB2**, where the mapping rules were divided into eight groups based on expert knowledge. The partitioning of mapping rules into groups in **TB2** resulted in a positive impact on the execution time of KG creation. Furthermore, the planner was able to find plans that were less costly in terms of execution time (measured in seconds) across all the engines studied. The highest savings were achieved when the mapping rules were initially partitioned by the experts, followed by executing the planner for each group of the partition.

RML Engine	TB1	TB2	Savings %
SDM-RDFizer	4,885.09 (s)	2,657.23 (s)	45.61%
SDM-RDFizer+Planner	2,931.054 (s)	1,535.9 (s)	47.60%
Morph-KGC	1,017.72 (s)	848.6 (s)	16.62%
Morph-KGC+Planner	781.5 (s)	592.4 (s)	24.2%
RMLMapper	Timed out (Five hours) 0% of Results	Timed out (Five hours) 19.37% of Results	19.37%
RMLMapper+Planner	Timed out (Five hours) 38.34% of Results	Timed out (Five hours) 57.49% of Results	19.15%

dependencies, and formats. Inter-class constraints, on the other hand, involve referential integrity, cardinality and connectivity, as well as mandatory and optional relationships among classes in the KG ontology. These constraints are specified using the W3C SHACL language and validated using Trav-SHACL [22]. Trav-SHACL is a SHACL engine capable of optimizing the validation process through the reordering of shapes in a shape schema and rewriting target and constraint queries. By validating these shapes, we are able to curate KG4C and address any existing ambiguities in collaboration with clinical data providers.

Federated Query Processing: DeTrusty [60] is a federated query engine able to execute queries against a federation of data sources accessible via SPARQL endpoints. First, DeTrusty maintains the description of KGs in a federation in terms of RDF Molecule Templates (RDF-MTs) or RDF classes and their properties existing in the KGs. DeTrusty receives a SPARQL query Q and generates the answers by combining data collected from several KGs. First, DeTrusty conduces the task of Source Selection and Query Decomposition to identify a query decomposition of a query Q. DeTrusty resorts to metadata encoded in RDF-MTs to find a solution to the problem of query decomposition and source selection, i.e., a decomposition of the initial query into the KGs. These selected KGs and the sub-queries are utilized to generate an execution plan. The Query Optimizer employs various optimization strategies to identify an efficient query plan, i.e., a plan of Q over the relevant KGs that produces all the answers but in the minimal time. DeTrusty also implements different physical operators (e.g., agjoin and adjoin [1]) and merges the data in a continuous fashion. DeTrusty is evaluated on nine queries defined over the KG4C federation. They comprise between 14 and 22 triple patterns and up to 21 joins. These queries retrieve up

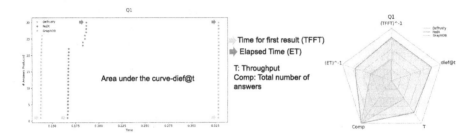

Fig. 4. Explanation of Result Plots.

to 41 results but produce a large number of intermediate results; the queries are in Appendix A. DeTrusty is compared with GraphDB[8] and FedEx (RDF4J)[9]. GraphDB is a commercial SPARQL engine able to evaluate federated queries, while FedEx (RDF4J) is a federated query engine which extends the work of Schwarte et al. [65]. We report the following metrics: **a)** *Execution Time (ET)*: Elapsed time between the submission of a query to a query engine and the generation of the answers. Time corresponds to absolute wall-clock system time as reported by the Python time.time() function. **b)** *Completeness (Comp)*: Query result percentage with respect to the query answer cardinality. **c)** T: Throughput quantifies the rate at which a query engine generates the query answers. **d)** *TFFT*: Time for the first tuple. **e)** *dief@t*: A measurement for the continuous efficiency of an engine in the first t time units of its execution [2]; it computes AUC (area-under-the-curve) of the answer distribution until time t; a higher value means the query engine has a steadier answer production. The average execution time and standard deviation over 10 runs are reported in Appendix B. KG4C is accessible via a SPARQL endpoint implemented in Virtuoso 7.20.3229 configured to use up to 64 GiB. The experiments are executed on an Ubuntu 18.04.6 LTS 64 bit machine with an Intel® Xeon® W-2133 CPU (six physical cores, twelve threads), and 64 GiB DDR4 RAM. A timeout of 10 min is considered, and all the caches are flushed between the execution of two queries to ensure reproducibility. The block or "paginating" is configured to 10,000 answers. The continuous behavior of DeTrusty is reported in radar plots. These plots encode the results of the inverse times for the first answer (*TFFT*) and the total execution time *ET*. Additionally, *Comp* and T represent the percentage of completeness of the produced query answer and throughput. Lastly, *dief@t* measures an engine's steady generation of a query answer.

Figure 4 compares the traces of the generation of query answers (figure left-side) and how these traces are utilized to create the radar plots (figure right-side). As observed, these plots enable the characterization of the continuous behavior of these engines. The comparison of the behavior of the three studied engines is presented in Fig. 5. All three engines yield the same number of results for the

---

[8] https://www.ontotext.com/products/graphdb/.
[9] https://rdf4j.org/documentation/programming/federation/.

nine queries. However, GraphDB timed out in Q6, while DeTrusty outperforms GraphDB in all metrics. On the other hand, FedEx (RDF4J) performs better than GraphDB and is competitive to DeTrusty in queries Q5 and Q8. The reason behind this is that these two queries are both selective and generate a relatively small number of results: 492 for Q5 and 1,124 for Q8. However, the remaining queries produce a larger number of intermediate results (e.g., 5,989 for Q4 and 16,811 for Q9), require access to three KGs, or involve more than two instances of the general predicate owl:sameAs. Notably, the predicate owl:sameAs is present in all the KGs, and FedEx (RDF4J) lacks the capability to decide where the triple patterns of this predicate should be evaluated. As a result, FedEx (RDF4J) poses the execution of these triple patterns to all the KGs. In contrast, DeTrusty utilizes source descriptions and can identify that these triple patterns need to be evaluated solely against KG4C. Another significant advantage of DeTrusty lies in its physical operators, which can incrementally generate query answers. Consequently, the DeTrusty query decomposition and planning methods not only select the minimal number of KGs required to produce the answer for each sub-query but also produce results as soon as they are computed. These combined factors enable DeTrusty to outperform these two engines.

Symbolic Learning: KGs represent knowledge about entities and their properties and relations in the form of factual statements. Inductive learning models resort to machine learning techniques to uncover rules that describe patterns among the represented entities and potential predictions. Specifically, the method proposed by Galarraga et al. [25] extracts Horn clauses of the form:

$$\text{hasBio(?a, ALK)} \Leftarrow \text{ageCategory(?a,young), sex(?a,female)}$$

This rule is extracted from the portion of the KG4C that includes only lung cancer patients; it indicates that "Lung cancer patients represented by the variable ?a who are young and females are also likely ALK translocated." Each mined rule is associated with the metrics of *Support, Head Coverage, Partial Completeness Assumption (PCA)*, and *F-score (F-PCA-HC)*. **a)** *Support*: a rule r counts the number of instantiations of the *Body* that are also true in the *Head*. The support of the rule in the example is 46, which corresponds to the number of instantiations from the KG– illustrated in Fig. 1– which make the *Body* and *Head* of the rule true. **b)** *Head Coverage*: corresponds to fraction of the instantiation of the *Head* that are true in the *Body*. In the running example, there are 318 instantiations of the *Head* and only 46 are true in the *Body* and *Head*, i.e., *Head Coverage (HC)* is equal to $\frac{46}{318} = 0.14$. **c)** *PCA*: corresponds to the fraction of the possible instantiations of the *Head* out of the true instantiations of the *Body*. In the running example, 52 out of the 206 lung cancer patients who are young and females are also ALK translocated, i.e., $PCA = \frac{52}{206} = 0.25$. **d)** *F-PCA-HC*: represents the harmonic mean of *Head Coverage* and *PCA*, i.e., 0.17 is the *F-PCA-HC* of the running example rule.

AMIE, the algorithm proposed by Galarraga et al. [25], was used to mine rules from a portion of KG4C that includes 17,000 lung cancer patients, their demographic characteristics (e.g., age, sex, smoking habits), previous cancer his-

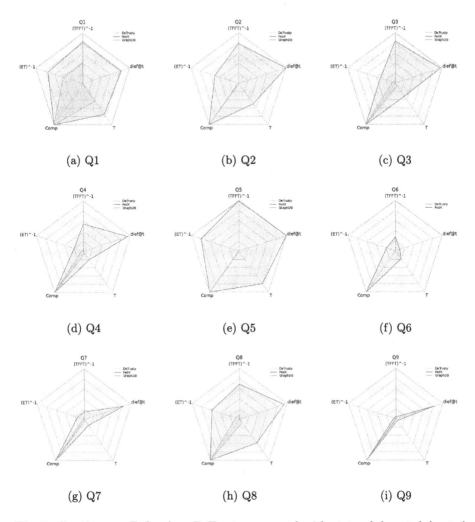

(a) Q1   (b) Q2   (c) Q3

(d) Q4   (e) Q5   (f) Q6

(g) Q7   (h) Q8   (i) Q9

Fig. 5. Continuous Behavior. DeTrusty compared with state-of-the-art federated query engines: FedEx (RDF4J) and GraphDB in terms of: $(TFFF)^{-1}$ - inverse time for the first answer produced, $(ET)^{-1}$ - inverse query execution time, Comp - number of answers, T - throughput, and $dief@t$ - continuous efficiency at time t. DeTrusty produces the first query answer ahead of the other engines and finalizes before. DeTrusty continuously produces the answers.

tory, biomarkers (e.g., EGFR, ALK, PDL1), cancer stages, and familial cancer antecedents. AMIE was configured to mine Horn rules with maximum three predicates, and with minimum HC of 0.01 and minimum PCA confidence of 0.1. AMIE was executed on an Ubuntu 18.04.6 LTS 64 bit machine with an Intel® Xeon® W-2133 CPU (six physical cores, twelve threads), and 64 GiB DDR4 RAM. In total, 17,767 Horn rules were mined in 1 h and 45 min. The PCA values range from 0.1 to 1.0, while the *HC* took values in the range of 0.01 to 1.0. We will discuss some rules in the next section.

4 Data Analytics Over Knowledge Graphs

The Horn rules extracted by AMIE over KG4C are used to support oncologists in the screening and identification of people with high risk of developing lung cancer. Mined rules are analyzed in terms of biomarkers and previous cancer, and ranked based on *PCA*, *HC*, and *F-PCA-HC*. The rules were discussed with two oncologists; the most relevant ones are reported in Table 3. The extracted rules reveal the following patterns:

Biomarkers

- PDL1 positive lung cancer patients are smokers and males. These patients are in advanced cancer stages (i.e., IIIC or IVB). Similar results are observed in the population studied in the analysis reported in [23,76].
- ALK translocation and EGFR, smoking habits (i.e., never-smokers) and sex (i.e., females) seem to be relevant features of the lung cancer patients who are positive for these two biomarkers. Sweis et al. [67] report that alterations of these genes were observed in with higher percentages in light- or never-smokers, while Ha et al. [31] present results that support that lung cancer in non-smoker Asian females is most often in EGFR patients.
- KRAS patients are likely to be smokers, either current or former; this observation is consistent with the statement that KRAS mutations are more commonly present in Western lung cancer patients who are smokers [58].

History of Previous Cancers

- Based on the mined rules, lung cancer patients who previously suffered breast cancer are commonly females, diagnosed in advanced stages, PDL1 positive, and never-smokers. These statements are supported by Nobel et al. [54] and Wennstig et al. [73] which report that lung cancer is the most common cancer in breast cancer survivors. Additionally, Gatalica et al. [26] indicate that a significant portion of metastatic tumors (e.g., breast cancer) to the lungs are PDL1 expression. Lasty, Wang et al. [72] report the results of second primary lung cancer of female breast cancer patients, who are the majority diagnosed in advanced stages (stage II or more).
- According to the mined rules: lung cancer patients who previously, suffered prostate cancer or head and neck, are mostly males and smokers. Pagedar et al. [55] report that lung cancer is the most commonly identified second primary malignancy in head and neck cancer, with a male predominance and current or former smokers. Zhang et al. [77] report cumulative incidence of lung as a second cancer of prostate cancer patients.

Table 3. Mined Horn Rules. Exemplary mined and statements Reported in the Literature and their Relationship with the Analysis Outcomes.

Statement	Exemplary Mined Horn Rules
PDL1 positive lung cancer patients are likely smokers and males [23,76]. These patients are in advanced cancer stages (i.e., IIIC or IVB) [78].	hasBio(?a, PDL1) ⇐ hasStage(?a,IIIC), sex(?a,male)
ALK translocation and EGFR, smoking habits (i.e., never-smokers) and sex (i.e., females) seem to be relevant features of the lung cancer patients who are positive for these two biomarkers [31].	hasBio(?a, EGFR) ⇐ hasStage(?a,IV), smoker(?a,nonSmoker)
KRAS patients are likely to be males, old, and smokers, either current or former [58].	hasBio(?a, KRAS) ⇐ age(?a,old), smoker(?a,heavySmoker)
Lung cancer patients who previously suffered breast cancer according to he mined patterns, these patients are females, diagnosed in advanced stages, with familial cancer history, PDL1 positive, and never-smokers [54,73].	hasPreviousCancer(?a,breast) ⇐ sex(?a,female), stage(?a,IV)

5   Related Work

The literature extensively addresses the problem of devising data integration frameworks [32]. The mediator and wrapper architecture proposed by Wiederhold [74] and the data integration system approach presented by Lenzerini [45] represent seminal works. The database and semantic web communities have extensively contributed to the problem and defined a large number of frameworks and formalisms [18,28,32,33,50,51]. Despite the large piece of work, semantically integrating data collected from multimodal and heterogeneous data sources still represents a challenge. In this work, we aim to present a pipeline that integrates data based on the declarative specification of a data integration system, and generates, as a result, a knowledge graph.

The problem of defining mapping languages that enable to specify correspondences across heterogeneous data sources has also received considerable attention [14,17,37,43,56]. The scientific community has also extensively treated the problem of federated query processing [1,19–21,30,49,61,65,69]. This paper provides evidence of the importance of planning and query optimization to ensure efficient executions of KG creation pipelines and query processing.

Knowledge graphs have been accepted as data structures to represent data and knowledge using a graph data model [34]. Several approaches have been proposed for transforming textual and unstructured data into KGs [8,39,63,64]. Additionally, a large piece of work has been devoted to frameworks for KG creation [5,6,17,36]. Lastly, several computational frameworks have been defined to perform symbolic learning over KGs [24,25,42,47]. This paper illustrates how KGs can be used in the context of healthcare to uncover patterns that contribute to a better understanding of a disease, e.g., lung cancer.

6   Conclusions and Future Directions

We tackle the problem of healthcare data analysis and position KGs as expressive data structures that enable the integration of data collected from heterogeneous data sources, and data integration systems as frameworks for declaratively defining KGs. We present various data management techniques to transform heterogeneous data sources into a unified KG. Moreover, the use of KGs is illustrated in the context of lung cancer, where symbolic learning methods enable to mine patterns– in the form of Horn clauses– the represent reported statements from the literature. Thus, we aim to extend the repertoire of techniques that can be utilized to support data analytics on the healthcare area, as well as encourage the community to empower existing approaches for semantic data integration and query processing to efficiently create, manage, and explore KGs.

Despite years of research, there are still several challenges that still require attention. They include: **i)** Approaches for multimodal data integration (e.g., images, unstructured and omics data); **ii)** Explainable and interpretable methods for KG creation, curation, management, and exploration; **iii)** Methods to ensure privacy and sovereignty; **iv)** Techniques to mine patterns of causation over KGs; and **v)** Hybrid approaches that allow for the integration of human and machine intelligence towards more effective and efficient knowledge management and discovery. These grand challenges are part of our future research agenda.

Acknowledgement. This work has been supported by the EU H2020 RIA project CLARIFY (GA No. 875160). Maria-Esther Vidal is partially supported by Leibniz Association in the program "Leibniz Best Minds: Programme for Women Professors", project TrustKG-Transforming Data in Trustable Insights with grant P99/2020.

A   Federated SPARQL Queries

```
PREFIX KG4CE: <http://research.tib.eu/clarify2020/entity/>
PREFIX KG4CV: <http://research.tib.eu/clarify2020/vocab/>
PREFIX rdf: <http://www.w3.org/1999/02/22-rdf-syntax-ns#>
PREFIX owl: <http://www.w3.org/2002/07/owl#>

SELECT DISTINCT ?o1 ?do1 ?externaldo1
WHERE {
  ?patient rdf:type KG4CV:LCPatient .
  ?patient KG4CV:hasBio KG4CE:ALK .
  ?patient KG4CV:hasSmokingHabit KG4CE:NonSmoker .
  ?patient KG4CV:sex KG4CE:Female .
  ?patient KG4CV:age ?age .
  FILTER (?age < 51)
  ?patient KG4CV:hasTreatmentEpisode ?o .
  ?o rdf:type KG4CV:TreatmentEpisode .
  ?o KG4CV:hasTreatmentType ?hasSch .
  ?hasSch rdf:type KG4CV:Chemotherapy .
  ?hasSch KG4CV:hasDrugSchema ?schema .
  ?schema KG4CV:hasDrug1 ?o1 .
  ?o1 KG4CV:hasCUIAnnotation ?do1 .
  ?do1 rdf:type KG4CV:Annotation .
  ?do1 owl:sameAs ?externaldo1.
}
```

Listing 1: **Query 1**. Drugs that are part of at least one chemotherapy schema for female non-smoker lung cancer patients, and their external identifiers. The patients should be who are ALK translocated. Number of Triple Patterns: 14. Number of Sources: 1. Number of Results: 30.

```
PREFIX KG4CE: <http://research.tib.eu/clarify2020/entity/>
PREFIX KG4CV: <http://research.tib.eu/clarify2020/vocab/>
PREFIX rdf: <http://www.w3.org/1999/02/22-rdf-syntax-ns#>
PREFIX dbp: <http://dbpedia.org/property/>
PREFIX owl: <http://www.w3.org/2002/07/owl#>

SELECT DISTINCT ?o5 ?excretion ?metabolism ?routes
WHERE {
  ?patient rdf:type KG4CV:LCPatient .
  ?patient KG4CV:hasBio KG4CE:ALK .
  ?patient KG4CV:hasSmokingHabit KG4CE:NonSmoker .
  ?patient KG4CV:sex KG4CE:Female .
  ?patient KG4CV:age ?age .
  FILTER (?age < 51)
  ?patient KG4CV:hasTreatmentEpisode ?o .
  ?o rdf:type KG4CV:TreatmentEpisode .
  ?o KG4CV:hasTreatmentType ?hasSch .
  ?hasSch rdf:type KG4CV:Chemotherapy .
  ?hasSch KG4CV:hasDrugSchema ?schema .
  ?schema KG4CV:hasDrug1 ?o1 .
  ?o1 KG4CV:hasCUIAnnotation ?o4 .
  ?o4 rdf:type KG4CV:Annotation .
  ?o4 rdf:type KG4CV:Annotation .
  ?o4 owl:sameAs ?o5 .
  ?o5 dbp:excretion ?excretion .
  ?o5 dbp:metabolism ?metabolism .
  ?o5 dbp:routesOfAdministration ?routes
}
```

Listing 2: **Query 2**. Excretion, metabolism, and routes of administration of drugs that are part of at least one chemotherapy schema for female non-smoker lung cancer patients who are ALK translocated. Number of Triple Patterns: 18. Number of Sources: 2. Number of Results: 7.

```
PREFIX KG4CE: <http://research.tib.eu/clarify2020/entity/>
PREFIX KG4CV: <http://research.tib.eu/clarify2020/vocab/>
PREFIX rdf: <http://www.w3.org/1999/02/22-rdf-syntax-ns#>
PREFIX wdt: <http://www.wikidata.org/prop/direct/>
PREFIX dbp: <http://dbpedia.org/property/>
PREFIX owl: <http://www.w3.org/2002/07/owl#>

SELECT DISTINCT ?o1 ?o4 ?o5 ?idDrug ?activeIngredient ?mass
WHERE {
  ?patient rdf:type KG4CV:LCPatient .
  ?patient KG4CV:hasBio KG4CE:EGFR .
  ?patient KG4CV:hasSmokingHabit KG4CE:NonSmoker .
  ?patient KG4CV:sex KG4CE:Female .
  ?patient KG4CV:hasTreatmentEpisode ?o .
  ?o rdf:type KG4CV:TreatmentEpisode .
  ?o KG4CV:hasTreatmentType ?hasSch .
  ?hasSch rdf:type KG4CV:Chemotherapy .
  ?hasSch KG4CV:hasDrugSchema ?schema .
  ?schema KG4CV:hasDrug1 ?o1 .
  ?o1 KG4CV:hasCUIAnnotation ?o4 .
  ?o4 rdf:type KG4CV:Annotation .
  ?o4 owl:sameAs ?o5 .
  ?o5 wdt:P592 ?idDrug .
  ?o5 wdt:P3780 ?activeIngredient .
  ?o5 wdt:P2067 ?mass .
}
```

Listing 3: **Query 3**. ChEMBL ID, active ingredients, and mass of drugs that are part of at least one chemotherapy schema for female non-smoker lung cancer patients who are EGFR positive. Number of Triple Patterns: 16. Number of Sources: 2. Number of Results: 30.

```
PREFIX KG4CE: <http://research.tib.eu/clarify2020/entity/>
PREFIX KG4CV: <http://research.tib.eu/clarify2020/vocab/>
PREFIX rdf: <http://www.w3.org/1999/02/22-rdf-syntax-ns#>
PREFIX wdt: <http://www.wikidata.org/prop/direct/>
PREFIX dbp: <http://dbpedia.org/property/>
PREFIX owl: <http://www.w3.org/2002/07/owl#>

SELECT DISTINCT ?o1 ?o4 ?o5 ?idDrug ?activeIngredient ?mass
WHERE {
    ?patient rdf:type KG4CV:LCPatient .
    ?patient KG4CV:hasBio KG4CE:PDL1 .
    ?patient KG4CV:sex KG4CE:Male .
    ?patient KG4CV:hasTreatmentEpisode ?o .
    ?o rdf:type KG4CV:TreatmentEpisode .
    ?o KG4CV:hasTreatmentType ?hasSch .
    ?hasSch rdf:type KG4CV:Chemotherapy .
    ?hasSch KG4CV:hasDrugSchema ?schema .
    ?schema KG4CV:hasDrug1 ?o1 .
    ?o1 KG4CV:hasCUIAnnotation ?o4 .
    ?o4 rdf:type KG4CV:Annotation .
    ?o4 owl:sameAs ?o5 .
    ?o5 wdt:P592 ?idDrug .
    ?o5 wdt:P3780 ?activeIngredient .
    ?o5 wdt:P2067 ?mass .
}
```

Listing 4: **Query 4.** ChEMBL ID, active ingredients, and mass of drugs that are part of at least one chemotherapy schema for male lung cancer patients who are PDL1 positive. Number of Triple Patterns: 15. Number of Sources: 2. Number of Results: 41.

```
PREFIX KG4CE: <http://research.tib.eu/clarify2020/entity/>
PREFIX KG4CV: <http://research.tib.eu/clarify2020/vocab/>
PREFIX rdf: <http://www.w3.org/1999/02/22-rdf-syntax-ns#>
PREFIX wdt: <http://www.wikidata.org/prop/direct/>
PREFIX dbp: <http://dbpedia.org/property/>
PREFIX owl: <http://www.w3.org/2002/07/owl#>
SELECT DISTINCT ?o1 ?o4 ?o5 ?idDrug ?activeIngredient ?mass
WHERE {
  ?patient rdf:type KG4CV:LCPatient .
  ?patient KG4CV:hasBio KG4CE:EGFR .
  ?patient KG4CV:hasSmokingHabit KG4CE:NonSmoker .
  ?patient KG4CV:sex KG4CE:Male .
  ?patient KG4CV:age ?age.
  FILTER (?age > 51) .
  ?patient KG4CV:hasTreatmentEpisode ?o .
  ?o rdf:type KG4CV:TreatmentEpisode .
  ?o KG4CV:hasTreatmentType ?hasSch .
  ?hasSch rdf:type KG4CV:Chemotherapy .
  ?hasSch KG4CV:hasDrugSchema ?schema .
  ?schema KG4CV:hasDrug1 ?o1 .
  ?o1 KG4CV:hasCUIAnnotation ?o4 .
  ?o4 rdf:type KG4CV:Annotation .
  ?o4 owl:sameAs ?o5 .
  ?o5 wdt:P592 ?idDrug .
  ?o5 wdt:P3780 ?activeIngredient .
  ?o5 wdt:P2067 ?mass .
}
```

Listing 5: **Query 5.** ChEMBL ID, active ingredients, and mass of drugs that are part of at least one chemotherapy schema for male non-smoker lung cancer patients who are EGFR positive. Number of Triple Patterns: 17. Number of Sources: 2. Number of Results: 22.

```
PREFIX KG4CE: <http://research.tib.eu/clarify2020/entity/>
PREFIX KG4CV: <http://research.tib.eu/clarify2020/vocab/>
PREFIX rdf: <http://www.w3.org/1999/02/22-rdf-syntax-ns#>
PREFIX wdt: <http://www.wikidata.org/prop/direct/>
PREFIX dbp: <http://dbpedia.org/property/>
PREFIX owl: <http://www.w3.org/2002/07/owl#>
SELECT DISTINCT ?o1 ?o4 ?o5 ?excretion ?metabolism ?routes ?o6 ?idDrug
  ?activeIngredient ?mass
WHERE {
  ?patient rdf:type KG4CV:LCPatient .
  ?patient KG4CV:hasBio KG4CE:EGFR .
  ?patient KG4CV:hasSmokingHabit KG4CE:NonSmoker .
  ?patient KG4CV:sex KG4CE:Male .
  ?patient KG4CV:age ?age .
  FILTER (?age > 51) .
  ?patient KG4CV:hasTreatmentEpisode ?o .
  ?o rdf:type KG4CV:TreatmentEpisode .
  ?o KG4CV:hasTreatmentType ?hasSch .
  ?hasSch rdf:type KG4CV:Chemotherapy .
  ?hasSch KG4CV:hasDrugSchema ?schema .
  ?schema KG4CV:hasDrug1 ?o1 .
  ?o1 KG4CV:hasCUIAnnotation ?o4 .
  ?o4 rdf:type KG4CV:Annotation .
  ?o4 owl:sameAs ?o5 .
  ?o4 rdf:type KG4CV:Annotation .
  ?o4 owl:sameAs ?o6 .
  ?o6 dbp:excretion ?excretion .
  ?o6 dbp:metabolism ?metabolism .
  ?o6 dbp:routesOfAdministration ?routes .
  ?o5 wdt:P592 ?idDrug .
  ?o5 wdt:P3780 ?activeIngredient .
  ?o5 wdt:P2067 ?mass .
}
```

Listing 6: **Query 6**. ChEMBL ID, active ingredients, mass, excretion, metabolism, and routes of administration of drugs that are part of at least one chemotherapy schema for male non-smoker lung cancer patients who are EGFR positive. Number of Triple Patterns: 22. Number of Sources: 3. Number of Results: 19.

```
PREFIX KG4CE: <http://research.tib.eu/clarify2020/entity/>
PREFIX KG4CV: <http://research.tib.eu/clarify2020/vocab/>
PREFIX rdf: <http://www.w3.org/1999/02/22-rdf-syntax-ns#>
PREFIX wdt: <http://www.wikidata.org/prop/direct/>
PREFIX dbp: <http://dbpedia.org/property/>
PREFIX owl: <http://www.w3.org/2002/07/owl#>
SELECT DISTINCT ?o1 ?o4 ?o5 ?excretion ?metabolism ?routes ?o6 ?idDrug
  ?activeIngredient ?mass
WHERE {
  ?patient rdf:type KG4CV:LCPatient .
  ?patient KG4CV:hasBio KG4CE:EGFR .
  ?patient KG4CV:hasSmokingHabit KG4CE:NonSmoker .
  ?patient KG4CV:sex KG4CE:Female .
  ?patient KG4CV:hasTreatmentEpisode ?o .
  ?o rdf:type KG4CV:TreatmentEpisode .
  ?o KG4CV:hasTreatmentType ?hasSch .
  ?hasSch KG4CV:hasDrugSchema ?schema .
  ?hasSch rdf:type KG4CV:Chemotherapy .
  ?schema KG4CV:hasDrug1 ?o1 .
  ?o1 KG4CV:hasCUIAnnotation ?o4 .
  ?o4 rdf:type KG4CV:Annotation .
  ?o4 owl:sameAs ?o5 .
  ?o4 owl:sameAs ?o6 .
  ?o6 dbp:excretion ?excretion .
  ?o6 dbp:metabolism ?metabolism .
  ?o6 dbp:routesOfAdministration ?routes .
  ?o5 wdt:P592 ?idDrug .
  ?o5 wdt:P3780 ?activeIngredient .
  ?o5 wdt:P2067 ?mass .
}
```

Listing 7: **Query 7**. ChEMBL ID, active ingredients, mass, excretion, metabolism, and routes of administration of drugs that are part of at least one chemotherapy schema for female non-smoker lung cancer patients who are EGFR positive. Number of Triple Patterns: 20. Number of Sources: 3. Number of Results: 24.

```
PREFIX KG4CE: <http://research.tib.eu/clarify2020/entity/>
PREFIX KG4CV: <http://research.tib.eu/clarify2020/vocab/>
PREFIX rdf: <http://www.w3.org/1999/02/22-rdf-syntax-ns#>
PREFIX wdt: <http://www.wikidata.org/prop/direct/>
PREFIX dbp: <http://dbpedia.org/property/>
PREFIX owl: <http://www.w3.org/2002/07/owl#>
SELECT DISTINCT ?o1 ?o4 ?o5 ?excretion ?metabolism ?routes ?o6 ?idDrug
   ?activeIngredient ?mass
WHERE {
   ?patient rdf:type KG4CV:LCPatient .
   ?patient KG4CV:hasBio KG4CE:ALK .
   ?patient KG4CV:sex KG4CE:Male .
   ?patient KG4CV:hasTreatmentEpisode ?o .
   ?o rdf:type KG4CV:TreatmentEpisode .
   ?o KG4CV:hasTreatmentType ?hasSch .
   ?hasSch rdf:type KG4CV:Chemotherapy .
   ?hasSch KG4CV:hasDrugSchema ?schema .
   ?schema KG4CV:hasDrug1 ?o1 .
   ?o1 KG4CV:hasCUIAnnotation ?o4 .
   ?o4 rdf:type KG4CV:Annotation .
   ?o4 owl:sameAs ?o5 .
   ?o4 owl:sameAs ?o6 .
   ?o6 dbp:excretion ?excretation .
   ?o6 dbp:metabolism ?metabolism .
   ?o6 dbp:routesOfAdministration ?routes .
   ?o5  wdt:P592 ?idDrug .
   ?o5  wdt:P3780 ?activeIngredient .
   ?o5  wdt:P2067 ?mass .
}
```

Listing 8: **Query 8**. ChEMBL ID, active ingredients, mass, excretion, metabolism, and routes of administration of drugs that are part of at least one chemotherapy schema for male non-smoker lung cancer patients who are ALK translocated. Number of Triple Patterns: 19. Number of Sources: 3. Number of Results: 19.

```
PREFIX KG4CE: <http://research.tib.eu/clarify2020/entity/>
PREFIX KG4CV: <http://research.tib.eu/clarify2020/vocab/>
PREFIX rdf: <http://www.w3.org/1999/02/22-rdf-syntax-ns#>
PREFIX wdt: <http://www.wikidata.org/prop/direct/>
PREFIX dbp: <http://dbpedia.org/property/>
PREFIX owl: <http://www.w3.org/2002/07/owl#>
SELECT DISTINCT ?o1 ?o4 ?o5 ?excretion ?metabolism ?routes ?o6 ?idDrug
  ?activeIngredient ?mass
WHERE {
  ?patient rdf:type KG4CV:LCPatient .
  ?patient KG4CV:hasBio KG4CE:PDL1 .
  ?patient KG4CV:hasTreatmentEpisode ?o .
  ?o rdf:type KG4CV:TreatmentEpisode .
  ?o KG4CV:hasTreatmentType ?hasSch .
  ?hasSch rdf:type KG4CV:Chemotherapy .
  ?hasSch KG4CV:hasDrugSchema ?schema .
  ?schema KG4CV:hasDrug1 ?o1 .
  ?o1 KG4CV:hasCUIAnnotation ?o4 .
  ?o4 rdf:type KG4CV:Annotation .
  ?o4 owl:sameAs ?o5 .
  ?o4 owl:sameAs ?o6 .
  ?o6 dbp:excretion ?excretion .
  ?o6 dbp:metabolism ?metabolism .
  ?o6 dbp:routesOfAdministration ?routes .
  ?o5 wdt:P592 ?idDrug .
  ?o5 wdt:P3780 ?activeIngredient .
  ?o5 wdt:P2067 ?mass .
}
```

Listing 9: **Query 9**. ChEMBL ID, active ingredients, mass, excretion, metabolism, and routes of administration of drugs that are part of at least one chemotherapy schema for lung cancer patients who are PDL1 positive. Number of Triple Patterns: 18. Number of Sources: 3. Number of Results: 39.

B   Results Federated Query Engines

See Table 4.

Table 4. Execution Times of Federated Query Engines. *avg* is the average execution time of the query over 10 runs. *stdev* reports the standard deviation observed across the 10 runs. FedX (RDF4J) outperforms GraphDB in all queries. GraphDB times out for query Q6. DeTrusty has the best performance of all three engines in all nine queries. Additionally, the query execution time of DeTrusty is the most stable one as can be seen by the low standard deviation.

Query	DeTrusty		FedX (RDF4J)		GraphDB	
	avg	stdev	avg	stdev	avg	stdev
Q1	**0.16081**	0.00785	0.21581	0.02132	0.40062	0.04567
Q2	**0.27681**	0.01681	0.82318	0.03297	55.79853	1.29189
Q3	**0.58540**	0.04146	1.90763	0.20965	8.90954	0.43137
Q4	**1.06983**	0.07654	4.85232	0.08124	24.07908	0.08725
Q5	**0.45300**	0.08519	0.66207	0.03722	96.81955	2.64521
Q6	**0.81906**	0.08835	3.89900	0.09721	*timed out*	
Q7	**2.47035**	0.02817	17.63632	0.44751	86.98863	0.59556
Q8	**2.33876**	0.01786	4.34825	0.34618	28.66448	0.57074
Q9	**2.53045**	0.03086	56.85676	0.20893	326.44522	2.48396

References

1. Acosta, M., Vidal, M.-E., Lampo, T., Castillo, J., Ruckhaus, E.: ANAPSID: an adaptive query processing engine for SPARQL endpoints. In: Aroyo, L., et al. (eds.) ISWC 2011. LNCS, vol. 7031, pp. 18–34. Springer, Heidelberg (2011). https://doi.org/10.1007/978-3-642-25073-6_2

2. Acosta, M., Vidal, M.-E., Sure-Vetter, Y.: Diefficiency metrics: measuring the continuous efficiency of query processing approaches. In: d'Amato, C., et al. (eds.) ISWC 2017. LNCS, vol. 10588, pp. 3–19. Springer, Cham (2017). https://doi.org/10.1007/978-3-319-68204-4_1

3. Aisopos, F., et al.: Knowledge graphs for enhancing transparency in health data ecosystems. Semant. Web **14**(5), 943–976 (2023). https://doi.org/10.3233/SW-223294

4. Angell, R., Monath, N., Mohan, S., Yadav, N., McCallum, A.: Clustering-based inference for biomedical entity linking. In: Proceedings of the 2021 Conference of the North American Chapter of the Association for Computational Linguistics: Human Language Technologies, pp. 2598–2608 (2021). https://doi.org/10.18653/v1/2021.naacl-main.205

5. Arenas-Guerrero, J., et al.: Knowledge graph construction with r2rml and rml: an ETL system-based overview. In: Proceedings of the 2nd International Workshop on Knowledge Graph Construction Co-located with 18th Extended Semantic Web Conference (ESWC 2021), Online, 6 June 2021. CEUR Workshop Proceedings, vol. 2873. CEUR-WS.org (2021). https://ceur-ws.org/Vol-2873/paper11.pdf

6. Arenas-Guerrero, J., Chaves-Fraga, D., Toledo, J., Pérez, M.S., Corcho, O.: Morph-KGC: scalable knowledge graph materialization with mapping partitions. Semant. Web (2022). https://doi.org/10.3233/SW-223135

7. Badenes-Olmedo, C., et al.: Drugs4Covid: drug-driven knowledge exploitation based on scientific publications. CoRR abs/2012.01953 (2020)

8. Barroca, J., Shivkumar, A., Ferreira, B.Q., Sherkhonov, E., Faria, J.: Enriching a fashion knowledge graph from product textual descriptions. arXiv preprint arXiv:2206.01087 (2022)

9. Beer, A., Brunet, M., Srivastava, V., Vidal, M.E.: Leibniz data manager - a research data management system. In: Groth, P., et al. (eds.) ESWC 2022. LNCS, vol. 13384, pp. 73–77. Springer, Cham (2022). https://doi.org/10.1007/978-3-031-11609-4_14

10. Benítez-Andrades, J.A., García-Ordás, M.T., Russo, M., Sakor, A., Fernandes, L.D., Vidal, M.E.: Empowering machine learning models with contextual knowledge for enhancing the detection of eating disorders in social media posts. Semant. Web **14**(5), 873–892 (2023). https://doi.org/10.3233/SW-223269

11. Chandak, P., Huang, K., Zitnik, M.: Building a knowledge graph to enable precision medicine. Sci. Data **10**(67) (2023). https://doi.org/10.1038/s41597-023-01960-3

12. Collarana, D., Galkin, M., Ribón, I.T., Lange, C., Vidal, M.E., Auer, S.: Semantic data integration for knowledge graph construction at query time. In: 11th IEEE International Conference on Semantic Computing, ICSC 2017, pp. 109–116 (2017). https://doi.org/10.1109/ICSC.2017.85

13. Collarana, D., Galkin, M., Traverso-Ribón, I., Vidal, M.E., Lange, C., Auer, S.: MINTE: semantically integrating RDF graphs. In: Proceedings of the 7th International Conference on Web Intelligence, Mining and Semantics (2017). https://doi.org/10.1145/3102254.3102280

14. Das, S., Sundara, S., Cyganiak, R.: R2RML: RDB to RDF Mapping Language, W3C Recommendation 27 September 2012. W3C (2012). http://www.w3.org/TR/r2rml/

15. Dimou, A.: Chapter 4 creation of knowledge graphs. In: Janev, V., Graux, D., Jabeen, H., Sallinger, E. (eds.) Knowledge Graphs and Big Data Processing. LNCS, vol. 12072, pp. 59–72. Springer, Cham (2020). https://doi.org/10.1007/978-3-030-53199-7_4

16. Dimou, A., Nies, T.D., Verborgh, R., Mannens, E., de Walle, R.V.: Automated metadata generation for linked data generation and publishing workflows. In: Proceedings of the Workshop on Linked Data on the Web, LDOW 2016, Co-located with 25th International World Wide Web Conference (WWW 2016). CEUR Workshop Proceedings, vol. 1593. CEUR-WS.org (2016)

17. Dimou, A., Sande, M.V., Colpaert, P., Verborgh, R., Mannens, E., de Walle, R.V.: RML: a generic language for integrated RDF mappings of heterogeneous data. In: Proceedings of the Workshop on Linked Data on the Web co-located with the 23rd International World Wide Web Conference (WWW 2014), Seoul, Korea, 8 April 2014. CEUR Workshop Proceedings, vol. 1184. CEUR-WS.org (2014). https://ceur-ws.org/Vol-1184/ldow2014_paper_01.pdf

18. Doan, A., Halevy, A.Y., Ives, Z.G.: Principles of Data Integration. Morgan Kaufmann, Burlington (2012). http://research.cs.wisc.edu/dibook/

19. Endris, K.M., Galkin, M., Lytra, I., Mami, M.N., Vidal, M.-E., Auer, S.: Querying interlinked data by bridging RDF molecule templates. In: Hameurlain, A., Wagner, R., Benslimane, D., Damiani, E., Grosky, W.I. (eds.) Transactions on Large-Scale Data- and Knowledge-Centered Systems XXXIX. LNCS, vol. 11310, pp. 1–42. Springer, Heidelberg (2018). https://doi.org/10.1007/978-3-662-58415-6_1

20. Endris, K.M., Rohde, P.D., Vidal, M.-E., Auer, S.: Ontario: federated query processing against a semantic data lake. In: Hartmann, S., Küng, J., Chakravarthy, S., Anderst-Kotsis, G., Tjoa, A.M., Khalil, I. (eds.) DEXA 2019. LNCS, vol. 11706, pp. 379–395. Springer, Cham (2019). https://doi.org/10.1007/978-3-030-27615-7_29

21. Endris, K.M., Vidal, M.-E., Graux, D.: Chapter 5 federated query processing. In: Janev, V., Graux, D., Jabeen, H., Sallinger, E. (eds.) Knowledge Graphs and Big Data Processing. LNCS, vol. 12072, pp. 73–86. Springer, Cham (2020). https://doi.org/10.1007/978-3-030-53199-7_5

22. Figuera, M., Rohde, P.D., Vidal, M.E.: Trav-SHACL: efficiently validating networks of SHACL constraints. In: The Web Conference, pp. 3337–3348. ACM, New York, NY, USA (2021). https://doi.org/10.1145/3442381.3449877

23. Fu, F., Deng, C., Sun, W., et al.: Distribution and concordance of PD-L1 expression by routine 22C3 assays in East-Asian patients with non-small cell lung cancer. Respir. Res. **23**(302) (2022). https://doi.org/10.1186/s12931-022-02201-8

24. Galárraga, L., Teflioudi, C., Hose, K., Suchanek, F.: Fast rule mining in ontological knowledge bases with AMIE+. VLDB J. (2015). https://hal-imt.archives-ouvertes.fr/hal-01699866

25. Galárraga, L.A., Teflioudi, C., Hose, K., Suchanek, F.M.: AMIE: association rule mining under incomplete evidence in ontological knowledge bases. In: 22nd International World Wide Web Conference, WWW '13, Rio de Janeiro, Brazil, 13–17 May 2013, pp. 413–422. International World Wide Web Conferences Steering Committee/ACM (2013). https://doi.org/10.1145/2488388.2488425

26. Gatalica, Z., Senarathne, J., Vranic, S.: PD-L1 expression patterns in the metastatic tumors to the lung: a comparative study with the primary non-small cell lung cancer. Ann. Oncol. **28**(suppl_2), ii52 (2017). https://doi.org/10.1093/annonc/mdx094.003

27. Geisler, S., et al.: Knowledge-driven data ecosystems toward data transparency. ACM J. Data Inf. Qual. **14**(1), 3:1–3:12 (2022). https://doi.org/10.1145/3467022, https://doi.org/10.1145/3467022

28. Golshan, B., Halevy, A.Y., Mihaila, G.A., Tan, W.: Data integration: after the teenage years. In: Proceedings of the 36th ACM SIGMOD-SIGACT-SIGAI Symposium on Principles of Database Systems, PODS 2017, Chicago, IL, USA, 14–19 May 2017, pp. 101–106 (2017). https://doi.org/10.1145/3034786.3056124

29. Gries, D., Schneider, F.B.: A Logical Approach to Discrete Math. Texts and Monographs in Computer Science, Springer, Heidelberg (1993). https://doi.org/10.1007/978-1-4757-3837-7

30. Gu, Z., Corcoglioniti, F., Lanti, D., Mosca, A., Xiao, G., Xiong, J., Calvanese, D.: A systematic overview of data federation systems. Semant. Web, 1–59 (2022)

31. Ha, S., et al.: Lung cancer in never-smoker Asian females is driven by oncogenic mutations, most often involving EGFR. Oncotarget **10**(7) (2015). https://doi.org/10.18632/oncotarget.2925

32. Halevy, A.Y.: Information integration. In: Liu, L., Özsu, M.T. (eds.) Encyclopedia of Database Systems, 2nd edn. Springer, New York (2018). https://doi.org/10.1007/978-1-4614-8265-9_1069

33. Halevy, A.Y., Rajaraman, A., Ordille, J.J.: Data integration: the teenage years. In: Proceedings of the 32nd International Conference on Very Large Data Bases, Seoul, Korea, 12–15 September 2006, pp. 9–16 (2006)

34. Hogan, A., et al.: Knowledge Graphs. Synthesis Lectures on Data, Semantics, and Knowledge, Morgan & Claypool Publishers, San Rafael (2021). https://doi.org/10.2200/S01125ED1V01Y202109DSK022

35. Hulsen, T., et al.: From big data to precision medicine. Front. Med. **6** (2019). https://doi.org/10.3389/fmed.2019.00034, https://www.frontiersin.org/articles/10.3389/fmed.2019.00034

36. Iglesias, E., Jozashoori, S., Chaves-Fraga, D., Collarana, D., Vidal, M.E.: SDM-RDFizer: an RML interpreter for the efficient creation of rdf knowledge graphs. In: CIKM '20: The 29th ACM International Conference on Information and Knowledge Management, Virtual Event, Ireland, 19–23 October 2020, pp. 3039–3046. ACM (2020). https://doi.org/10.1145/3340531.3412881

37. Iglesias, E., Jozashoori, S., Vidal, M.E.: Scaling up knowledge graph creation to large and heterogeneous data sources. J. Web Semant. **75**, 100755 (2023). https://doi.org/10.1016/j.websem.2022.100755

38. Janev, V., et al.: Responsible knowledge management in energy data ecosystems. Energies **15**(11) (2022). https://doi.org/10.3390/en15113973

39. Jozashoori, S., Sakor, A., Iglesias, E., Vidal, M.E.: EABlock: a declarative entity alignment block for knowledge graph creation pipelines. In: SAC '22: The 37th ACM/SIGAPP Symposium on Applied Computing, Virtual Event, 25–29 April 2022, pp. 1908–1916. ACM (2022). https://doi.org/10.1145/3477314.3507132

40. Kilicoglu, H., Shin, D., Fiszman, M., Rosemblat, G., Rindflesch, T.: SemMedDB: a PubMed-scale repository of biomedical semantic predications. Bioinformatics **28**(23) (2012). https://doi.org/10.1093/bioinformatics/bts591

41. Krithara, A., et al.: iASiS: towards heterogeneous big data analysis for personalized medicine. In: 32nd IEEE International Symposium on Computer-Based Medical Systems. CBMS 2019, pp. 106–111. IEEE (2019). https://doi.org/10.1109/CBMS.2019.00032

42. Lajus, J., Galárraga, L., Suchanek, F.: Fast and exact rule mining with AMIE 3. In: Harth, A., et al. (eds.) ESWC 2020. LNCS, vol. 12123, pp. 36–52. Springer, Cham (2020). https://doi.org/10.1007/978-3-030-49461-2_3

43. Lefrançois, M., Zimmermann, A., Bakerally, N.: A SPARQL extension for generating rdf from heterogeneous formats. In: Blomqvist, E., Maynard, D., Gangemi, A., Hoekstra, R., Hitzler, P., Hartig, O. (eds.) ESWC 2017. LNCS, vol. 10249, pp. 35–50. Springer, Cham (2017). https://doi.org/10.1007/978-3-319-58068-5_3

44. Lehmann, J., et al.: DBpedia - a large-scale, multilingual knowledge base extracted from Wikipedia. Semant. Web J. (2015). https://doi.org/10.3233/SW-140134

45. Lenzerini, M.: Data integration: a theoretical perspective. In: Proceedings of the Twenty-First ACM SIGACT-SIGMOD-SIGART Symposium on Principles of Database Systems, 3–5 June, Madison, Wisconsin, USA, pp. 233–246. ACM (2002). https://doi.org/10.1145/543613.543644

46. Li, J., et al.: Biocreative v CDR task corpus: a resource for chemical disease relation extraction. Database (Oxford) 2016 (2016). https://doi.org/10.1093/database/baw068

47. Meilicke, C., Chekol, M.W., Ruffinelli, D., Stuckenschmidt, H.: Anytime bottom-up rule learning for knowledge graph completion. In: Proceedings of the Twenty-Eighth International Joint Conference on Artificial Intelligence, IJCAI 2019, Macao, China, 10–16 August 2019, pp. 3137–3143. ijcai.org (2019). https://doi.org/10.24963/ijcai.2019/435

48. Mohan, S., Li, D.: MedMentions: a large biomedical corpus annotated with UMLS concepts. In: Automated Knowledge Base Construction (AKBC) (2019). https://doi.org/10.24432/C5G59C

49. Montoya, G., Vidal, M.-E., Corcho, O., Ruckhaus, E., Buil-Aranda, C.: Benchmarking federated SPARQL query engines: are existing testbeds enough? In:

Cudré-Mauroux, P., et al. (eds.) ISWC 2012. LNCS, vol. 7650, pp. 313–324. Springer, Heidelberg (2012). https://doi.org/10.1007/978-3-642-35173-0_21

50. Mountantonakis, M.: Large scale services for connecting and integrating hundreds of linked datasets. SIGWEB Newsl. **2021**(Autumn), 3:1–3:4 (2021). https://doi.org/10.1145/3494825.3494828

51. Mountantonakis, M., Tzitzikas, Y.: Large-scale semantic integration of linked data: a survey. ACM Comput. Surv. **52**(5), 103:1–103:40 (2019). https://doi.org/10.1145/3345551

52. Namici, M., Giacomo, G.D.: Comparing query answering in OBDA tools over W3C-compliant specifications. In: Proceedings of the 31st International Workshop on Description Logics Co-located with 16th International Conference on Principles of Knowledge Representation and Reasoning (KR 2018), Tempe, Arizona, USA, 27–29 October 2018. CEUR Workshop Proceedings, vol. 2211. CEUR-WS.org (2018). https://ceur-ws.org/Vol-2211/paper-25.pdf

53. Neumann, M., King, D., Beltagy, I., Ammar, W.: ScispaCy: fast and robust models for biomedical natural language processing. In: Proceedings of the 18th BioNLP Workshop and Shared Task, pp. 319–327. Association for Computational Linguistics, Florence, Italy, August 2019. https://doi.org/10.18653/v1/w19-5034

54. Nobel, T.B., et al.: Primary lung cancer in women after previous breast cancer. BJS Open **5**(6), zrab115 (2022). https://doi.org/10.1093/bjsopen/zrab115

55. Pagedar, N.A., Jayawardena, A., Charlton, M.E., Hoffman, H.T.: Second primary lung cancer after head and neck cancer: Implications for screening computed tomography. Ann. Otol. Rhinol. Laryngol. **124**(10), 765–769 (2015). https://doi.org/10.1177/0003489415582259

56. Poggi, A., Lembo, D., Calvanese, D., Giacomo, G.D., Lenzerini, M., Rosati, R.: Linking data to ontologies. J. Data Semant. **10**, 133–173 (2008). https://doi.org/10.1007/978-3-540-77688-8_5

57. Ravi, M.P.K., Singh, K., Mulang, I.O., Shekarpour, S., Hoffart, J., Lehmann, J.: CHOLAN: a modular approach for neural entity linking on Wikipedia and Wikidata. In: Proceedings of the 16th Conference of the European Chapter of the Association for Computational Linguistics: Main Volume, pp. 504–514 (2021). https://doi.org/10.18653/v1/2021.eacl-main.40

58. Reck, M., Carbone, D.P., Garassino, M., Barlesi, F.: Targeting KRAS in non-small-cell lung cancer: recent progress and new approaches. Ann. Oncol. **32**(9), 1101–1110 (2021). https://doi.org/10.1016/j.annonc.2021.06.001

59. Rivas, A., Collarana, D., Torrente, M., Vidal, M.E.: A neuro-symbolic system over knowledge graphs for link prediction. Semant. Web (2023). https://www.semantic-web-journal.net/system/files/swj3324.pdf

60. Rohde, P.D., Bechara, M., Avellino: DeTrusty v0.12.2 (2023). https://doi.org/10.5281/zenodo.8063472

61. Ruckhaus, E., Ruiz, E., Vidal, M.E.: Query evaluation and optimization in the semantic web. Theory Pract. Log. Program. **8**(3), 393–409 (2008). https://doi.org/10.1017/S1471068407003225

62. Sakor, A., et al.: Knowledge4COVID-19: a semantic-based approach for constructing a COVID-19 related knowledge graph from various sources and analyzing treatments' toxicities. J. Web Semant. **75**, 100760 (2023). https://doi.org/10.1016/j.websem.2022.100760

63. Sakor, A., et al.: Old is gold: linguistic driven approach for entity and relation linking of short text. In: Proceedings of the 2019 Conference of the North American

Chapter of the Association for Computational Linguistics: Human Language Technologies, NAACL-HLT 2019, Volume 1 (Long Papers), pp. 2336–2346. Association for Computational Linguistics (2019). https://doi.org/10.18653/v1/n19-1243

64. Sakor, A., Singh, K., Patel, A., Vidal, M.E.: Falcon 2.0: an entity and relation linking tool over Wikidata. In: CIKM '20: The 29th ACM International Conference on Information and Knowledge Management, Virtual Event, Ireland, 19–23 October 2020, pp. 3141–3148. ACM (2020). https://doi.org/10.1145/3340531.3412777

65. Schwarte, A., Haase, P., Hose, K., Schenkel, R., Schmidt, M.: FedX: a federation layer for distributed query processing on linked open data. In: Antoniou, G., et al. (eds.) ESWC 2011. LNCS, vol. 6644, pp. 481–486. Springer, Heidelberg (2011). https://doi.org/10.1007/978-3-642-21064-8_39

66. Steenwinckel, B., et al.: Facilitating the analysis of COVID-19 literature through a knowledge graph. In: The Semantic Web - ISWC 2020, pp. 344–357 (2020). https://doi.org/10.1007/978-3-030-62466-8_22

67. Sweis, R., Thomas, S., Bank, B., Fishkin, P., Mooney, C., Salgia, R.: Concurrent EGFR mutation and ALK translocation in non-small cell lung cancer. Cureus **8**(2) (2016). https://doi.org/10.7759/cureus.513

68. Torrente, M., et al.: An artificial intelligence-based tool for data analysis and prognosis in cancer patients: results from the clarify study. Cancers **14**(16) (2022). https://doi.org/10.3390/cancers14164041

69. Vidal, M.E., Castillo, S., Acosta, M., Montoya, G., Palma, G.: On the selection of SPARQL endpoints to efficiently execute federated SPARQL queries. Trans. Large Scale Data Knowl. Centered Syst. **25**, 109–149 (2016). https://doi.org/10.1007/978-3-662-49534-6_4

70. Vidal, M.E., Endris, K.M., Jazashoori, S., Sakor, A., Rivas, A.: Transforming heterogeneous data into knowledge for personalized treatments - a use case. Datenbank-Spektrum **19**(2), 95–106 (2019). https://doi.org/10.1007/s13222-019-00312-z

71. Vrandečić, D., Krötzsch, M.: Wikidata: a free collaborative knowledgebase. Commun. ACM (2014). https://doi.org/10.1145/2629489

72. Wang, R., et al.: Second primary lung cancer after breast cancer: a population-based study of 6,269 women. Front. Oncol. **8**, 427 (2018). https://doi.org/10.3389/fonc.2018.00427

73. Wennstig, A.K., et al.: Risk of primary lung cancer after adjuvant radiotherapy in breast cancer-a large population-based study. NPJ Breast Cancer **7**(1), 71 (2021). https://doi.org/10.1038/s41523-021-00280-2

74. Wiederhold, G.: Mediators in the Architecture of Future Information Systems. IEEE Comput. **25**(3), 38–49 (1992)

75. Wu, B., Knoblock, C.A.: An iterative approach to synthesize data transformation programs. In: Proceedings of the 24th International Joint Conference on Artificial Intelligence (IJCAI) (2015)

76. Wu, X., et al.: PD-L1 expression correlation with metabolic parameters of FDG PET/CT and clinicopathological characteristics in non-small cell lung cancer. EJNMMI Res. **19**(1) (2020). https://doi.org/10.1186/s13550-020-00639-9

77. Zhang, H., Yu, A., Baran, A., Messing, E.: Risk of second cancer among young prostate cancer survivors. Radiat. Oncol. J. **39**(2), 91–98 (2021)

78. Zhao, Y., et al.: Prognostic significance of PD-L1 in advanced non-small cell lung carcinoma. Medicine (Baltimore) (2020). https://doi.org/10.1097/MD.0000000000023172

From Database Repairs to Causality
in Databases and Beyond

Leopoldo Bertossi[(✉)]

SKEMA Business School, Montreal, Canada
leopoldo.bertossi@skema.edu

Abstract. We describe some recent approaches to score-based explanations for query answers in databases. The focus is on work done by the author and collaborators. Special emphasis is placed on the use of counterfactual reasoning for score specification and computation. Several examples that illustrate the flexibility of these methods are shown.

1 Introduction

In data management one wants *explanations* for certain results. For example, for query results from databases. Explanations, that may come in different forms, have been the subject of philosophical enquires for a long time, but, closer to our discipline, they appear under different forms in model-based diagnosis and in causality as developed in artificial intelligence.

In the last few years, explanations that are based on *numerical scores* assigned to elements of a model that may contribute to an outcome have become popular. These scores attempt to capture the degree of contribution of those components to an outcome, e.g. answering questions like these: What is the contribution of this tuple to the answer to this query?

Different scores have been proposed in the literature, and some that have a relatively older history have been applied. Among the latter we find the general *responsibility score* as found in *actual causality* [11, 14]. For a particular kind of application, one has to define the right causality setting, and then apply the responsibility measure to the participating variables (see [15] for an updated treatment of causal responsibility).

In data management, responsibility has been used to quantify the strength of a tuple as a cause for a query result [4, 23] (see Sect. 3.1). The *responsibility score*, *Resp*, is based on the notions of *counterfactual intervention* as appearing in actual causality. More specifically, (potential) executions of *counterfactual interventions* on a *structural logico-probabilistic model* [14] are investigated, with the purpose of answering hypothetical questions of the form: *What would happen if we change ...?*.

Database repairs are commonly used to define and obtain semantically correct query answers from a database that may fail to satisfy a given set of integrity constraints (ICs) [3]. A connection between repairs and actual causality in DBs has been used to obtain complexity results and algorithms for the latter [4] (see Sect. 5).

L. Bertossi—Member of the Millennium Inst. for Foundational Research on Data (IMFD, Chile).

© Springer-Verlag GmbH Germany, part of Springer Nature 2023
A. Hameurlain et al. (Eds.): *TLDKS LIV*, LNCS 14160, pp. 119–131, 2023.
https://doi.org/10.1007/978-3-662-68014-8_5

The *Causal Effect* score is also based on causality, mainly for *observational studies* [16,26,29]. It has been applied in data management in [30] (see Sect. 3.2).

The Shapley value, as found in *coalition game theory* [31], has been used for the same purpose [18,19]. Defining the right game function, the *Shapley value* assigned to a player reflects its contribution to the wealth function. The Shapley value, which is firmly established in game theory, and is also used in several other areas [28,31]. The main idea is that *several tuples together*, much like players in a coalition game, are necessary to produce a query result. Some may contribute more than others to the *wealth distribution function* (or simply, game function), which in this case becomes the query result, namely 1 or 0 if the query is Boolean, or a number if we have an aggregation query. This use of Shapley value was developed in [18,19] (see Sect. 6).

In this article we survey some of the recent advances on the use and computation of the above mentioned score-based explanations for query answering in databases. This is not intended to be an exhaustive survey of the area. Instead, it is heavily influenced by our latest research. To introduce the concepts and techniques we will use mostly examples, trying to convey the main intuitions and issues.

This paper is structured as follows. In Sect. 2, we provide some preliminaries on databases. In Sect. 3, we introduce causality in databases and the responsibility score, and also the causal effect score. In Sect. 4, we show the connection between causality in databases and database repairs. In Sect. 5, we show how integrate ICs in the causality setting. In Sect. 6, we show how to use the Shapley value to provide explanation scores to database tuples in relation to a query result. In Sect. 7, we make some general remarks on relevant open problems.

2 Background

A relational schema $\mathcal{R}$ contains a domain of constants, $\mathcal{C}$, and a set of predicates of finite arities, $\mathcal{P}$. $\mathcal{R}$ gives rise to a language $\mathfrak{L}(\mathcal{R})$ of first-order (FO) predicate logic with built-in equality, $=$. Variables are usually denoted with $x, y, z, ...$; and finite sequences thereof with $\bar{x}, ...$; and constants with $a, b, c, ...$, etc. An *atom* is of the form $P(t_1, \ldots, t_n)$, with n-ary $P \in \mathcal{P}$ and $t_1, \ldots, t_n$ *terms*, i.e. constants, or variables. An atom is *ground* (a.k.a. a tuple) if it contains no variables. A database (instance), D, for $\mathcal{R}$ is a finite set of ground atoms; and it serves as an interpretation structure for $\mathfrak{L}(\mathcal{R})$.

A *conjunctive query* (CQ) is a FO formula, $\mathcal{Q}(\bar{x})$, of the form $\exists \bar{y} \, (P_1(\bar{x}_1) \wedge \cdots \wedge P_m(\bar{x}_m))$, with $P_i \in \mathcal{P}$, and (distinct) free variables $\bar{x} := (\bigcup \bar{x}_i) \smallsetminus \bar{y}$. If $\mathcal{Q}$ has n (free) variables, $\bar{c} \in \mathcal{C}^n$ is an *answer* to $\mathcal{Q}$ from D if $D \models \mathcal{Q}[\bar{c}]$, i.e. $Q[\bar{c}]$ is true in D when the variables in $\bar{x}$ are componentwise replaced by the values in $\bar{c}$. $\mathcal{Q}(D)$ denotes the set of answers to $\mathcal{Q}$ from D. $\mathcal{Q}$ is a *Boolean conjunctive query* (BCQ) when $\bar{x}$ is empty; and when *true* in D, $\mathcal{Q}(D) := \{true\}$. Otherwise, it is *false*, and $\mathcal{Q}(D) := \emptyset$. We will consider only conjunctive queries or disjunctions thereof.

We consider as integrity constraints (ICs), i.e. sentences of $\mathfrak{L}(\mathcal{R})$: (a) *denial constraints* (DCs), i.e. of the form $\kappa : \neg \exists \bar{x}(P_1(\bar{x}_1) \wedge \cdots \wedge P_m(\bar{x}_m))$, where $P_i \in \mathcal{P}$, and $\bar{x} = \bigcup \bar{x}_i$; and (b) *inclusion dependencies* (INDs), which are of the form $\forall \bar{x} \exists \bar{y}(P_1(\bar{x}) \rightarrow P_2(\bar{x}', \bar{y}))$, where $P_1, P_2 \in \mathcal{P}$, and $\bar{x}' \subseteq \bar{x}$. If an instance D does not satisfy the set Σ of ICs associated to the schema, we say that D is *inconsistent*, denoted with $D \not\models \Sigma$.

3   Causal Explanations in Databases

In data management we need to understand and compute *why* certain results are obtained or not, e.g. query answers, violations of semantic conditions, etc.; and we expect a database system to provide *explanations*.

3.1   Causal Responsibility

Here, we will consider *causality-based explanations* [23,24], which we will illustrate by means of an example.

Example 1.   Consider the database D, and the Boolean conjunctive query (BCQ)

R	A	B
	a	b
	c	d
	b	b

S	C
	a
	c
	b

Q : $\exists x \exists y (S(x) \wedge R(x,y) \wedge S(y))$, for which $D \models Q$ holds, i.e. the query is true in D. We ask about the causes for Q to be true.

A tuple $\tau \in D$ is *counterfactual cause* for Q (being true in D) if $D \models Q$ and $D \smallsetminus \{\tau\} \not\models Q$. In this example, $S(b)$ is a counterfactual cause for Q: If $S(b)$ is removed from D, Q is no longer true.

Removing a single tuple may not be enough to invalidate the query. Accordingly, a tuple $\tau \in D$ is an *actual cause* for Q if there is a *contingency set* $\Gamma \subseteq D$, such that τ is a counterfactual cause for Q in $D \smallsetminus \Gamma$. In this example, $R(a,b)$ is not a counterfactual cause for Q, but it is an actual cause with contingency set $\{R(b,b)\}$: If $R(b,b)$ is removed from D, Q is still true, but further removing $R(a,b)$ makes Q false. □

Notice that every counterfactual cause is also an actual cause, with empty contingent set. Actual causes that are not counterfactual causes need company to invalidate a query result. Now we ask how strong are tuples as actual causes. To answer this question, we appeal to the *responsibility* of an actual cause τ for Q [23], defined by:

$$Resp_D^{Q}(\tau) := \frac{1}{|\Gamma| + 1},$$

where $|\Gamma|$ is the size of a smallest contingency set, Γ, for τ, and 0, otherwise.

Example 2.   (ex. 1 cont.) The responsibility of $R(a,b)$ is $\frac{1}{2} = \frac{1}{1+1}$ (its several smallest contingency sets have all size 1). $R(b,b)$ and $S(a)$ are also actual causes with responsibility $\frac{1}{2}$; and $S(b)$ is actual (counterfactual) cause with responsibility $1 = \frac{1}{1+0}$.   □

High responsibility tuples provide more interesting explanations. Causes in this case are tuples that come with their responsibilities as "scores". All tuples can be seen as actual causes, but only those with non-zero responsibility score matter. Causality and responsibility in databases can be extended to the attribute-value level [4,6].

There are connections between database causality and *consistency-based diagnosis* and *abductive diagnosis*, that are two forms of *model-based diagnosis* [8,32]. There are also connections with *database repairs* [2,3]. These connections have led to complexity and algorithmic results for causality and responsibility [4,5] (see Sect. 4).

3.2 The Causal-Effect Score

Sometimes, as we will see right here below, responsibility does not provide intuitive or expected results, which led to the consideration of an alternative score, the *causal-effect score*. We show the issues and the score by means of an example.

Example 3.   Consider the database E that represents the graph below, and the Boolean query Q that is true in E if there is a path from a to b. Here, $E \models Q$. Tuples have global tuple identifiers (tids) in the left-most column, which is not essential, but convenient.

E	A	B
t_1	a	b
t_2	a	c
t_3	c	b
t_4	a	d
t_5	d	e
t_6	e	b

$Q:\ E(a,b)\ \vee$
$\exists x(E(a,x) \wedge E(x,b))\ \vee$
$\exists y \exists z(E(a,y) \wedge E(y,z) \wedge E(z,b))$

All tuples are actual causes since every tuple appears in a path from a to b. Also, all the tuples have the same causal responsibility, $\frac{1}{3}$, which may be counterintuitive, considering that t_1 provides a direct path from a to b.   □

In [30], the notion *causal effect* was introduced. It is based on three main ideas, namely, the transformation, for auxiliary purposes, of the database into a probabilistic database, the expected value of a query, and interventions on the lineage of the query [10,33]. The lineage of a query represents, by means of a propositional formula, all the ways in which the query can be true in terms of the potential database tuples, and their combinations. Here, "potential" refers to tuples that can be built with the database predicates and the database (finite) domain. These tuples may belong to the database at hand or not. For a given database, D, some of those atoms become true, and others false, which leads to the instantiation of the lineage (formula) o D.

Example 4.   Consider the database D below, and a BCQ.

R	A	B
	a	b
	a	c
	c	b

S	C
	b
	c

$Q:\ \exists x \exists y(R(x,y) \wedge S(y))$, which is true in D.

For the database D in our example, the lineage of the query instantiated on D is given by the propositional formula:

$$\Phi_Q(D) = (X_{R(a,b)} \wedge X_{S(b)}) \vee (X_{R(a,c)} \wedge X_{S(c)}) \vee (X_{R(c,b)} \wedge X_{S(b)}), \quad (1)$$

where X_τ is a propositional variable that is true iff $\tau \in D$. Here, $\Phi_Q(D)$ takes value 1 in D.

Now, for illustration, we want to quantify the contribution of tuple $S(b)$ to the query answer. For this purpose, we assign, uniformly and independently, probabilities to the tuples in D, obtaining a *probabilistic database* D^p [33]. Potential tuples outside D get probability 0.

R^p	A	B	prob
	a	b	$\frac{1}{2}$
	a	c	$\frac{1}{2}$
	c	b	$\frac{1}{2}$

S^p	C	prob
	b	$\frac{1}{2}$
	c	$\frac{1}{2}$

The X_τ's become independent, identically distributed Boolean random variables; and Q becomes a Boolean random variable. Accordingly, we can ask about the probability that Q takes the truth value 1 (or 0) when an *intervention* is performed on D.

Interventions are of the form $do(X = x)$, meaning making X take value x, with $x \in \{0, 1\}$, in the *structural model*, in this case, the lineage. That is, we ask, for $\{y, x\} \subseteq \{0, 1\}$, about the conditional probability $P(Q = y \mid do(X_\tau = x))$, i.e. conditioned to making X_τ false or true.

For example, with $do(X_{S(b)} = 0)$ and $do(X_{S(b)} = 1)$, the lineage in (1) becomes, resp., and abusing the notation a bit:

$$\Phi_Q(D \mid do(X_{S(b)} = 0)) := (X_{R(a,c)} \wedge X_{S(c)}).$$
$$\Phi_Q(D \mid do(X_{S(b)} = 1)) := X_{R(a,b)} \vee (X_{R(a,c)} \wedge X_{S(c)}) \vee X_{R(c,b)}.$$

On the basis of these lineages and D^p, when $X_{S(b)}$ is made false, the probability that the instantiated lineage becomes true in D^p is:

$$P(Q = 1 \mid do(X_{S(b)} = 0)) = P(X_{R(a,c)} = 1) \times P(X_{S(c)} = 1) = \frac{1}{4}.$$

When $X_{S(b)}$ is made true, the probability of the lineage being true in D^p is:

$$P(Q = 1 \mid do(X_{S(b)} = 1)) = P(X_{R(a,b)} \vee (X_{R(a,c)} \wedge X_{S(c)}) \vee X_{R(c,b)} = 1) = \frac{13}{16}.$$

The *causal effect* of a tuple τ is defined by:

$$\mathcal{CE}^{D,Q}(\tau) := \mathbb{E}(Q \mid do(X_\tau = 1)) - \mathbb{E}(Q \mid do(X_\tau = 0)).$$

In particular, using the probabilities computed so far:

$$\mathbb{E}(Q \mid do(X_{S(b)} = 0)) = P(Q = 1 \mid do(X_{S(b)} = 0)) = \frac{1}{4},$$
$$\mathbb{E}(Q \mid do(X_{S(b)} = 1)) = P(Q = 1 \mid do(X_{S(b)} = 1)) = \frac{13}{16}.$$

Then, the causal effect for the tuple $S(b)$ is: $\mathcal{CE}^{D,Q}(S(b)) = \frac{13}{16} - \frac{1}{4} = \frac{9}{16} > 0$, showing that the tuple is relevant for the query result, with a relevance score provided by the causal effect, of $\frac{9}{16}$. □

Let us now retake the initial example of this section.

Example 5. (ex. 3 cont.) The query has the lineage:

$$\Phi_{\mathcal{Q}}(D) = X_{t_1} \vee (X_{t_2} \wedge X_{t_3}) \vee (X_{t_4} \wedge X_{t_5} \wedge X_{t_6}).$$

It holds:

$$\mathcal{CE}^{D,\mathcal{Q}}(t_1) = 0.65625,$$
$$\mathcal{CE}^{D,\mathcal{Q}}(t_2) = \mathcal{CE}^{D,\mathcal{Q}}(t_3) = 0.21875,$$
$$\mathcal{CE}^{D,\mathcal{Q}}(t_4) = \mathcal{CE}^{D,\mathcal{Q}}(t_5) = \mathcal{CE}^{D,\mathcal{Q}}(t_6) = 0.09375.$$

The causal effects are different for different tuples, and the scores are much more intuitive than the responsibility scores. □

4   The Database Repair Connection

In this section we will first establish a useful connection between database repairs and causes as tuples in a database [2,3]. The notion of *repair* of a relational database was introduced in order to formalize the notion of *consistent query answering* (CQA), as shown in Fig. 1: If a database D is inconsistent in the sense that is does not satisfy a given set of integrity constraints, *ICs*, and a query $\mathcal{Q}$ is posed to D (left-hand side of Fig. 1), what are the meaningful, or consistent, answers to $\mathcal{Q}$ from D? They are sanctioned as those that hold (are returned as answers) from *all* the *repairs* of D. The repairs of D are consistent instances D' (over the same schema of D), i.e. $D' \models ICs$, and *minimally depart* from D (right-hand side of Fig. 1).

Notice that: (a) We have now a *possible-world* semantics for (consistent) query answering; and (b) we may use in principle any reasonable notion of distance between database instances, with each choice defining a particular *repair semantics*. In the rest of this section we will illustrate two classes of repairs, which have been used and investigated the most in the literature. Actually, repairs in general have got a life of their own, beyond consistent query answering.

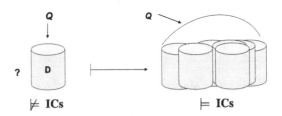

Fig. 1. Database repairs and consistent query answers

Example 6. Let us consider the following set of *denial constraints* (DCs) and a database D, whose relations (tables) are shown right here below. D is inconsistent, because it violates the DCs: it satisfies the joins that are prohibited by the DCs.

$$\neg\exists x\exists y(P(x)\wedge Q(x,y))$$
$$\neg\exists x\exists y(P(x)\wedge R(x,y))$$

P	A
	a
	e

Q	A	B
	a	b

R	A	C
	a	c

We want to repair the original instance by *deleting tuples* from relations. Notice that, for DCs, insertions of new tuple will not restore consistency. We could change (update) attribute values though, a possibility that has been investigated in [6].

Here we have two *subset repairs*, a.k.a. *S-repairs*. They are subset-maximal consistent subinstances of D: $D_1 = \{P(e), Q(a,b), R(a,c)\}$ and $D_2 = \{P(e), P(a)\}$. They are consistent, subinstances of D, and any proper superset of them (still contained in D) is inconsistent. (In general, we will represent database relations as set of tuples.)

We also have *cardinality repairs*, a.k.a. *C-repairs*. They are consistent subinstances of D that minimize the *number* of tuples by which they differ from D. That is, they are maximum-cardinality consistent subinstances. In this case, only D_1 is a C-repair. Every C-repair is an S-repair, but not necessarily the other way around. □

Let us now consider a BCQ

$$\mathcal{Q}: \exists\bar{x}(P_1(\bar{x}_1)\wedge\cdots\wedge P_m(\bar{x}_m)), \tag{2}$$

which we assume is true in a database D. It turns out that we can obtain the causes for $\mathcal{Q}$ to be true D, and their contingency sets from database repairs. In order to do this, notice that $\neg\mathcal{Q}$ becomes a DC

$$\kappa(\mathcal{Q}): \neg\exists\bar{x}(P_1(\bar{x}_1)\wedge\cdots\wedge P_m(\bar{x}_m)); \tag{3}$$

and that $\mathcal{Q}$ holds in D iff D is inconsistent w.r.t. $\kappa(\mathcal{Q})$.

It holds that S-repairs are associated to causes with minimal contingency sets, while C-repairs are associated to causes for $\mathcal{Q}$ with minimum contingency sets, and maximum responsibilities [4]. In fact, for a database tuple $\tau\in D$:

(a) τ is actual cause for $\mathcal{Q}$ with subset-minimal contingency set Γ iff $D\smallsetminus(\Gamma\cup\{\tau\})$ is an S-repair (w.r.t. $\kappa(\mathcal{Q})$), in which case, its responsibility is $\frac{1}{1+|\Gamma|}$.

(b) τ is actual cause with minimum-cardinality contingency set Γ iff $D\smallsetminus(\Gamma\cup\{\tau\})$ is C-repair, in which case, τ is a maximum-responsibility actual cause.

Conversely, repairs can be obtained from causes and their contingency sets [4]. These results can be extended to unions of BCQs (UBCQs), or equivalently, to sets of denial constraints.

One can exploit the connection between causes and repairs to understand the computational complexity of the former by leveraging existing results for the latter. Beyond the fact that computing or deciding actual causes can be done in polynomial time in data for CQs and UCQs [4,23], one can show that most computational problems related to responsibility are hard, because they are also hard for repairs, in particular, for C-repairs (all this in data complexity) [20]. In particular, one can prove [4]: (a) The *responsibility problem*, about deciding if a tuple has responsibility above a certain threshold, is *NP*-complete for UCQs. (b) Computing $Resp_D^{\mathcal{Q}}(\tau)$ is $FP^{NP(log(n))}$-complete for BCQs.

This the *functional*, non-decision, version of the responsibility problem. The complexity class involved is that of computational problems that use polynomial time with a logarithmic number of calls to an oracle in *NP*. (c) Deciding if a tuple τ is a most responsible cause is $P^{NP(log(n))}$-complete for BCQs. The complexity class is as the previous one, but for decision problems [1].

5   Causes Under Integrity Constraints

In this section we consider tuples as causes for query answering in the more general setting where databases are subject to integrity constraints (ICs). In this scenario, and in comparison with Sect. 3.1, not every intervention on the database is admissible, because the ICs have to be satisfied. As a consequence, the definitions of cause and responsibility have to be modified accordingly. We illustrate the issues by means of an example. More details can be found in [5,6].

We start assuming that a database D satisfies a set of ICs, Σ, i.e. $D \models \Sigma$. If we concentrate on BCQs, or more, generally on monotone queries, and consider causes at the tuple level, only instances obtained from D by interventions that are tuple deletions have to be considered; and they should satisfy the ICs. More precisely, for τ to be actual cause for $\mathcal{Q}$, with a contingency set Γ, it must hold [5]:

(a) $D \smallsetminus \Gamma \models \Sigma$, and $D \smallsetminus \Gamma \models \mathcal{Q}$.
(b) $D \smallsetminus (\Gamma \cup \{\tau\}) \models \Sigma$, and $D \smallsetminus (\Gamma \cup \{\tau\}) \not\models \mathcal{Q}$.

The *responsibility* of τ, denoted $Resp_{D,\Sigma}^{\mathcal{Q}}(\tau)$, is defined as in Sect. 3.1, through minimum-size contingency sets.

Example 7. Consider the database instance D below, initially without additional ICs.

Dep	DName	TStaff
t_1	Computing	John
t_2	Philosophy	Patrick
t_3	Math	Kevin

Course	CName	TStaff	DName
t_4	COM08	John	Computing
t_5	Math01	Kevin	Math
t_6	HIST02	Patrick	Philosophy
t_7	Math08	Eli	Math
t_8	COM01	John	Computing

Let us first consider the following open query:[1]

$$\mathcal{Q}(x)\colon \; \exists y \exists z (Dep(y,x) \wedge Course(z,x,y)). \tag{4}$$

In this case, we get answers other that *yes* or *no*. Actually, $\langle \text{John} \rangle \in \mathcal{Q}(D)$, the set of answers to $\mathcal{Q}$, and we look for causes for this particular answer. It holds: (a) t_1 is a counterfactual cause; (b) t_4 is actual cause with single minimal contingency set $\Gamma_1 = \{t_8\}$; (c) t_8 is actual cause with single minimal contingency set $\Gamma_2 = \{t_4\}$.

Let us now impose on D the *inclusion dependency* (IND):

---

[1] The fact that it is open is not particularly relevant, because we can instantiate the query with the answer, obtaining a Boolean query.

$$\psi: \quad \forall x \forall y \, (Dep(x,y) \rightarrow \exists u \, Course(u,y,x)), \tag{5}$$

which is satisfied by D. Now, t_4 t_8 are not actual causes anymore; and t_1 is still a counterfactual cause.

Let us now consider the query: $\mathcal{Q}_1(x)\colon \exists y \, Dep(y,x)$. Now, $\langle \text{John} \rangle \in \mathcal{Q}_1(D)$, and under the IND (5), we obtain the same causes as for Q, which is not surprising considering that $\mathcal{Q} \equiv_\psi \mathcal{Q}_1$, i.e. the two queries are logically equivalent under (5).

And now, consider the query: $\mathcal{Q}_2(x)\colon \exists y \exists z \, Course(z,x,y)$, for which $\langle \text{John} \rangle \in \mathcal{Q}_2(D)$. For this query we consider the two scenarios, with and without imposing the IND. Without imposing (5), t_4 and t_8 are the only actual causes, with contingency sets $\Gamma_1 = \{t_8\}$ and $\Gamma_2 = \{t_4\}$, resp.

However, imposing (5), t_4 and t_8 are still actual causes, but we lose their smallest contingency sets Γ_1 and Γ_2 we had before: $D \smallsetminus (\Gamma_1 \cup \{t_4\}) \not\models \psi$, $D \smallsetminus (\Gamma_2 \cup \{t_8\}) \not\models \psi$. Actually, the smallest contingency set for t_4 is $\Gamma_3 = \{t_8, t_1\}$; and for t_8, $\Gamma_4 = \{t_4, t_1\}$. We can see that under the IND, the responsibilities of t_4 and t_8 decrease:

$$Resp_D^{\mathcal{Q}_2(\text{John})}(t_4) = \frac{1}{2}, \text{ and } Resp_{D,\psi}^{\mathcal{Q}_2(\text{John})}(t_4) = \frac{1}{3}.$$

Tuple t_1 is not an actual cause, but it affects the responsibility of actual causes., $\square$

Some results about causality under ICs can be obtained [5]: (a) Causes are preserved under logical equivalence of queries under ICs, (b) Without ICs, deciding causality for BCQs is tractable, but their presence may make complexity grow. More precisely, there are a BCQ and an inclusion dependency for which deciding if a tuple is an actual cause is NP-complete in data.

6   The Shapley Value in Databases

The Shapley value was proposed in game theory by Lloyd Shapley in 1953 [31], to quantify the contribution of a player to a coalition game where players share a wealth function.[2] It has been applied in many disciplines. In particular, it has been investigated in computer science under *algorithmic game theory* [25], and it has been applied to many and different computational problems. The computation of the Shapley value is, in general, intractable. In many scenarios where it is applied its computation turns out to be $\#P$-hard [12, 13]. Here, the class $\#P$ contains the problems of *counting* the solutions for problems in NP. A typical problem in the class, actually, hard for the class, is $\#SAT$, about counting the number of satisfying assignments for a propositional formula. Clearly, this problem cannot be easier than SAT, because a solution for $\#SAT$ immediately gives a solution for SAT [1].

---

[2] The original paper and related ones on the Shapley value can be found in the book edited by Alvin Roth [28]. Shapley and Roth shared the Nobel Prize in Economic Sciences 2012.

Consider a set of players D, and a game function, $\mathcal{G} : \mathcal{P}(D) \to \mathbb{R}$, where $\mathcal{P}(D)$ the power set of D. The Shapley value of player p in D es defined by:

$$Shapley(D, \mathcal{G}, p) := \sum_{S \subseteq D \setminus \{p\}} \frac{|S|!(|D| - |S| - 1)!}{|D|!} (\mathcal{G}(S \cup \{p\}) - \mathcal{G}(S)). \quad (6)$$

Notice that here, $|S|!(|D| - |S| - 1)!$ is the number of permutations of D with all players in S coming first, then p, and then all the others. That is, this quantity is the expected contribution of player p under all possible additions of p to a partial random sequence of players followed by a random sequence of the rests of the players. Notice the counterfactual flavor, in that there is a comparison between what happens having p vs. not having it. The Shapley value is the only function that satisfies certain natural properties in relation to games. So, it is a result of a categorical set of axioms or conditions [28].

The Shapley value has been used in knowledge representation, to measure the degree of inconsistency of a propositional knowledge base [17]; in machine learning to provide explanations for the outcomes of classification models on the basis of numerical scores assigned to the participating feature values [21,22]. It has also been applied in data management to measure the contribution of a tuple to a query answer [18,19], which we briefly review in this section.

In databases, the players are tuples in a database D. We also have a Boolean query $\mathcal{Q}$, which becomes a game function, as follows: For $S \subseteq D$, i.e. a subinstance,

$$\mathcal{Q}(S) = \begin{cases} 1 & \text{if } S \models \mathcal{Q}, \\ 0 & \text{if } S \not\models \mathcal{Q}. \end{cases}$$

With these elements we can define the Shapley value of a tuple $\tau \in D$:

$$Shapley(D, \mathcal{Q}, \tau) := \sum_{S \subseteq D \setminus \{\tau\}} \frac{|S|!(|D| - |S| - 1)!}{|D|!} (\mathcal{Q}(S \cup \{\tau\}) - \mathcal{Q}(S)).$$

If the query is *monotone*, i.e. its set of answers never shrinks when new tuples are added to the database, which is the case of conjunctive queries (CQs), among others, the difference $\mathcal{Q}(S \cup \{\tau\}) - \mathcal{Q}(S)$ is always 1 or 0, and the average in the definition of the Shapley value returns a value between 0 and 1. This value quantifies the contribution of tuple τ to the query result. It was introduced and investigated in [18,19], for BCQs and some aggregate queries defined over CQs. We report on some of the findings in the rest of this section. The analysis has been extended to queries with negated atoms in CQs [27].

A main result obtained in [18,19] is about the complexity of computing this Shapley score. The following *Dichotomy Theorem* holds: For $\mathcal{Q}$ a BCQ without self-joins, if $\mathcal{Q}$ is *hierarchical*, then $Shapley(D, \mathcal{Q}, \tau)$ can be computed in polynomial-time (in the size of D); otherwise, the problem is #P-complete.

Here, $\mathcal{Q}$ is hierarchical if for every two existential variables x and y, it holds: (a) $Atoms(x) \subseteq Atoms(y)$, or $Atoms(y) \subseteq Atoms(x)$, or $Atoms(x) \cap Atoms(y) = \emptyset$. For example, $\mathcal{Q} : \exists x \exists y \exists z (R(x, y) \land S(x, z))$, for which $Atoms(x) = \{R(x, y),$

$S(x, z)\}$, $Atoms(y) = \{R(x, y)\}$, $Atoms(z) = \{S(x, z)\}$, is hierarchical. However, $\mathcal{Q}^{nh} : \exists x \exists y (R(x) \wedge S(x, y) \wedge T(y))$, for which $Atoms(x) = \{R(x), S(x, y)\}$, $Atoms(y) = \{S(x, y), T(y)\}$, is not hierarchical.

These are the same criteria for (in)tractability that apply to evaluation of BCQs over probabilistic databases [33]. However, the same proofs do not apply, at least not straightforwardly. The intractability result uses query $\mathcal{Q}^{nh}$ above, and a reduction from counting independent sets in a bipartite graph.

The dichotomy results can be extended to summation over CQs, with the same conditions and cases. This is because the Shapley value, as an expectation, is linear. Hardness extends to aggregates max, min, and avg over non-hierarchical queries.

For the hard cases, there is, as established in [18, 19], an *approximation result*: For every fixed BCQ $\mathcal{Q}$ (or summation over a CQ), there is a *multiplicative fully-polynomial randomized approximation scheme* (FPRAS) [1], A, with, for given ϵ and δ:

$$P(\tau \in D \mid \frac{Shapley(D, \mathcal{Q}, \tau)}{1 + \epsilon} \leq A(\tau, \epsilon, \delta) \leq (1 + \epsilon) Shapley(D, \mathcal{Q}, \tau)\}) \geq 1 - \delta.$$

A related and popular score, in coalition games and other areas, is the *Banzhaf Power Index*, which is similar to the Shapley value, but the order of players is ignored, by considering subsets of players rather than permutations thereof. It is defined by:

$$Banzhaf(D, \mathcal{Q}, \tau) := \frac{1}{2^{|D|-1}} \cdot \sum_{S \subseteq (D \setminus \{\tau\})} (\mathcal{Q}(S \cup \{\tau\}) - \mathcal{Q}(S)).$$

The Banzhaf-index is also difficult to compute; provably #P-hard in general. The results in [18, 19] carry over to this index when applied to query answering. In [18] it was proved that the causal-effect score of Sect. 3.2 coincides with the Banzhaf-index, which gives to the former an additional justification.

In [9], additional applications of the Shapley value in databases are described.

7   Final Remarks

Explainable data management and explainable AI (XAI) are effervescent areas of research. The relevance of explanations can only grow, as observed from- and due to the legislation and regulations that are being produced and enforced in relation to explainability, transparency and fairness of data management and AI/ML systems.

There are different approaches and methodologies in relation to explanations, with causality, counterfactuals and scores being prominent approaches that have a relevant role to play. Much research is still needed on the use of *contextual, semantic and domain knowledge*. Some approaches may be more appropriate in this direction, and we argue that declarative, logic-based specifications can be successfully exploited [7].

Still fundamental research is needed in relation to the notions of *explanation* and *interpretation*. An always present question is: *What is a good explanation?*. This is not a new question, and in AI (and other areas and disciplines) it has been investigated. In particular in AI, areas such as *diagnosis* and *causality* have much to contribute.

Now, in relation to *explanations scores*, there is still a question to be answered: *What are the desired properties of an explanation score?*. The question makes a lot of sense, and may not be beyond an answer. After all, the general Shapley value emerged from a list of *desiderata* in relation to coalition games, as the only measure that satisfies certain explicit properties [28,31]. Although the Shapley value is being used in XAI, in particular in its *Shap* incarnation, there could be a different and specific set of desired properties of explanation scores that could lead to a still undiscovered explanation score.

Acknowledgments. Part of this work was funded by ANID - Millennium Science Initiative Program - Code ICN17002; and NSERC-DG 2023-04650.

References

1. Arora, S., Barak, B.: Computational Complexity. Cambridge University Press, Cambridge (2009)
2. Arenas, M., Bertossi, L., Chomicki, J. Consistent query answers in inconsistent databases. In: Proceedings of the ACM PODS, pp. 68–79 (1999)
3. Bertossi, L.: Database repairing and consistent query answering. Synthesis Lectures in Data Management. Morgan & Claypool (2011)
4. Bertossi, L., Salimi, B.: From Causes for database queries to repairs and model-based diagnosis and back. Theory Comput. Syst. **61**(1), 191–232 (2017). https://doi.org/10.1007/s00224-016-9718-9
5. Bertossi, L., Salimi, B.: Causes for query answers from databases: datalog abduction, view-updates, and integrity constraints. Int. J. Approximate Reason. **90**, 226–252 (2017)
6. Bertossi, L.: Specifying and computing causes for query answers in databases via database repairs and repair programs. Knowl. Inf. Syst. **63**(1), 199–231 (2021)
7. Bertossi, L.: Declarative approaches to counterfactual explanations for classification. Theory Pract. Logic Program. **23**(3), 559–593 (2023). arXiv Paper 2011.07423
8. Bertossi, L.: Attribution-scores and causal counterfactuals as explanations in artificial intelligence. In: Reasoning Web: Causality, Explanations and Declarative Knowledge. Springer LNCS 13759 (2023). https://doi.org/10.1007/978-3-031-31414-8_1
9. Bertossi, L., Kimelfeld, B., Livshits, E., Monet, M.: The Shapley Value in Database Management. ACM SIGMOD Rec. **52**(2), 6–17 (2023)
10. Buneman, P., Khanna, S., Tan, W.C.: Why and where: a characterization of data provenance. Proceedings of ICDT, pp. 316–330 (2001)
11. Chockler, H., Halpern, J.: Responsibility and blame: a structural-model approach. J. Artif. Intell. Res. **22**, 93–115 (2004)
12. Deng, X., Papadimitriou, C.: On the complexity of cooperative solution concepts. Math. Oper. Res. **19**(2), 257–266 (1994)
13. Faigle, U., Kern, W.: The Shapley value for cooperative games under precedence constraints. Int. J. Game Theory **21**, 249–266 (1992)
14. Halpern, J., Pearl, J.: Causes and explanations: a structural-model approach. Part I: Causes British J. Philos. Sci. **56**(4), 843–887 (2005)
15. Halpern, J.Y.: A modification of the halpern-pearl definition of causality. In: Proceedings of IJCAI, pp. 3022–3033 (2015)
16. Holland, P.W.: Statistics and causal inference. J. Am. Statist. Assoc. **81**(396), 945–960 (1986)
17. Hunter, A., Konieczny, S.: On the measure of conflicts: Shapley inconsistency values. Artif. Intell. **174**(14), 1007–1026 (2010)

18. Livshits, E., Bertossi, L., Kimelfeld, B., Sebag, M.: The Shapley value of tuples in query answering. Logical Methods Comput. Sci. **17**(3), 22.1-22.33 (2021)
19. Livshits, E., Bertossi, L., Kimelfeld, B., Sebag, M.: Query games in databases. ACM SIG-MOD Rec. **50**(1), 78–85 (2021)
20. Lopatenko, A., Bertossi, L.: Complexity of consistent query answering in databases under cardinality-based and incremental repair semantics. In: Schwentick, T., Suciu, D. (eds.) ICDT 2007. LNCS, vol. 4353, pp. 179–193. Springer, Heidelberg (2006). https://doi.org/10.1007/11965893_13
21. Lundberg, S., et al.: From local explanations to global understanding with explainable AI for trees. Nat. Mach. Intell. **2**(1), 2522–5839 (2020)
22. Lundberg, S., Lee, S.: A unified approach to interpreting model predictions. In: Proceedings of Advances in Neural Information Processing Systems, pp. 4765–4774 (2017)
23. Meliou, A., Gatterbauer, W., Moore, K.F., Suciu, D.: The complexity of causality and responsibility for query answers and non-answers. In: Proceedings of VLDB, pp. 34–41 (2010)
24. Meliou, A., Gatterbauer, W., Halpern, J.Y., Koch, C., Moore, K.F., Suciu, D.: Causality in databases. IEEE Data Eng. Bull. **33**(3), 59–67 (2010)
25. Nisan, N., Roughgarden, T., Tardos, E., Vazirani, V.V. (eds.): Algorithmic Game Theory. Cambridge University Press (2007)
26. Pearl, J.: Causality: Models, Reasoning and Inference, 2nd edn. Cambridge University Press, Cambridge (2009)
27. Reshef, A., Kimelfeld, B., Livshits, E.: The impact of negation on the complexity of the Shapley value in conjunctive queries. In: Proceedings of PODS, pp. 285–297 (2020)
28. Roth, A.E. (ed.): The Shapley Value: Essays in Honor of Lloyd S. Cambridge University Press, Shapley (1988)
29. Rubin, D.B.: Estimating causal effects of treatments in randomized and nonrandomized studies. J. Educ. Psychol. **66**, 688–701 (1974)
30. Salimi, B., Bertossi, L., Suciu, D., Van den Broeck, G.: Quantifying causal effects on query answering in databases. In: Proceedings of the 8th USENIX Workshop on the Theory and Practice of Provenance (TaPP) (2016)
31. Shapley, L.S.: A value for n-person games. Contrib. Theory Games **2**(28), 307–317 (1953)
32. Struss, P.: Model-based problem solving. In: Handbook of Knowledge Representation, Chap. 4. Elsevier, pp. 395–465 (2008)
33. Suciu, D., Olteanu, D., Re, C., Koch, C.: Probabilistic Databases. Morgan & Claypool, Synthesis Lectures on Data Management (2011)

Author Index

© Springer-Verlag GmbH Germany, part of Springer Nature 2023
A. Hameurlain et al. (Eds.): *TLDKS LIV*, LNCS 14160, p. 133, 2023.
https://doi.org/10.1007/978-3-662-68014-8

Printed in the United States
by Baker & Taylor Publisher Services